Application of Neural Networks to Modelling and Control

The Institute of Measurement and Control was founded in 1944 as the Society of Instrument Technology and took its present name in 1968. It was incorporated by Royal Charter in 1975 with the object '...to promote for the public benefit by all available means the general advancement of the science and practice of measurement and control technology and its application'. The Institute of Measurement and Control provides routes to Engineering Council status as Chartered and Incorporated Engineers and Engineering Technicians. The Institute of Measurement and Control is a registered charity, number 269815, and is based at 87 Gower Street, London WC1E 6AA. Telephone (071) 387 4949, facsimile (071) 388 8431.

Application of Neural Networks to Modelling and Control

Edited by

G.F. Page, J.B. Gomm and D. Williams

School of Electrical and Electronic Engineering,
Liverpool John Moores University,
Liverpool,
UK

CHAPMAN & HALL
London · Glasgow · New York · Tokyo · Melbourne · Madras

Published by Chapman & Hall, 2–6 Boundary Row, London SE1 8HN

Chapman & Hall, 2–6 Boundary Row, London SE1 8HN, UK

Blackie Academic & Professional, Wester Cleddens Road, Bishopbriggs, Glasgow G64 2NZ, UK

Chapman & Hall Inc., 29 West 35th Street, New York NY10001, USA

Chapman & Hall Japan, Thomas Publishing Japan, Hirakawacho Nemoto Building, 6F, 1–7–11 Hirakawa-cho, Chiyoda-ku, Tokyo 102, Japan

Chapman & Hall Australia, Thomas Nelson Australia, 102 Dodds Street, South Melbourne, Victoria 3205, Australia

Chapman & Hall India, R. Seshadri, 32 Second Main Road, CIT East, Madras 600 035, India

First edition 1993

© 1993 G.F. Page, J.B. Gomm and D. Williams

Typeset in 10/12 Times by Interprint Limited, Malta
Printed in England by Clays Ltd, St Ives plc

ISBN 0 412 54760 0

Apart from any fair dealing for the purposes of research or private study, or criticism or review, as permitted under the UK Copyright Designs and Patents Act, 1988, this publication may not be reproduced, stored, or transmitted, in any form or by any means, without the prior permission in writing of the publishers, or in the case of reprographic reproduction only in accordance with the terms of the licences issued by the Copyright Licensing Agency in the UK, or in accordance with the terms of licences issued by the appropriate Reproduction Rights Organization outside the UK. Enquiries concerning reproduction outside the terms stated here should be sent to the pubishers at the London address printed on this page.

The publisher makes no representation, express or implied, with regard to the accuracy of the information contained in this book and cannot accept any legal responsibility or liability for any errors or omissions that may be made.

A catalogue record for this book is available from the British Library

Library of Congress Cataloging-in-Publication data.
Application of neural networks to modelling and control/edited by
  G.F. Page, J.B. Gomm, and D. Williams.—1st ed.
    p. cm.
  Includes bibliographical references and index.
  ISBN 0-412-54760-0
  1. Neural networks (Computer science) 2. Computer simulation.
3. Automatic control. I. Page, G.F. II. Gomm. J.B.
III. Williams, D.
QA76.87.A65 1993
003'.363—dc20   93-14878
                                                          CIP

Contents

Preface	ix
Contributors	xiii
1 Introduction to neural networks	1
J.B. Gomm, G.F. Page and D. Williams	
1.1 Abstract	1
1.2 Introduction	1
1.3 Artificial neuron model	3
1.4 Multi-layer perceptron	4
1.5 Hopfield network	5
1.6 Kohonen network	6
1.7 Conclusions	7
References	7
2 Neural networks for control: evolution, revolution or renaissance	9
P.J.G. Lisboa	
2.1 Abstract	9
2.2 Introduction	9
2.3 Historical background	10
2.4 Estimation	13
2.5 Control	18
2.5.1 Adaptive critic	19
2.5.2 Copy control	19
2.5.3 Predictive control	20
2.5.4 Differentiating the model	21
2.5.5 Model reference control	21
2.6 Conclusions	22
Acknowledgements	23
References	23

3 Identification of linear systems using recurrent neural networks 25
D.T. Pham and X. Liu
- 3.1 Abstract 25
- 3.2 Introduction 25
- 3.3 Basic Elman net 26
- 3.4 Dynamic systems modelling using Elman nets 27
  - 3.4.1 Modified Elman net 27
  - 3.4.2 Simulation of dynamic systems 27
  - 3.4.3 System models used in simulations 29
  - 3.4.4 Simulation results and discussions 29
- 3.5 Conclusion 34
- Acknowledgement 34
- References 34

4 Enhancing feedforward neural network training 35
C. Peel, M.J. Willis and M.T. Tham
- 4.1 Abstract 35
- 4.2 Introduction 35
- 4.3 Feedforward artificial neural networks 37
- 4.4 Principal component analysis 39
- 4.5 Application of principal component analysis to neural networks 40
  - 4.5.1 Optimum topology selection 42
- 4.6 Discussion 43
- 4.7 Self-tuning/adaptive networks 44
- 4.8 Performance evaluation 44
  - 4.8.1 Estimation of product conversion in a continuous stirred tank reactor (CSTR) 45
  - 4.8.2 Estimation of melt flow index (MFI) in industrial polymerization 48
- 4.9 Concluding remarks 50
- Acknowledgements 51
- References 51

5 Estimation of state variables of a fermentation process via Kalman filter and neural network 53
D. Tsaptsinos, N.A. Jalel and J.R. Leigh
- 5.1 Abstract 53
- 5.2 Introduction 53
- 5.3 Modelling the fermentation process using an identification technique 55
  - 5.3.1 The AR model for the fermentation process 55
  - 5.3.2 Kalman filter estimator 56

5.3.3 Simulation results	56
5.3.4 Non-linear model of the process	59
5.3.5 Non-linear results	60
5.4 Modelling and state estimation using artificial neural networks	61
5.4.1 The feedforward neural network	61
5.4.2 The preparation phase: I	62
5.4.3 The preparation phase: II	62
5.4.4 Stop-learning criterion	63
5.4.5 Using correlation analysis to reduce the network topology	63
5.4.6 Post-pruning processes I and II	64
5.4.7 Improving the performance	66
5.4.8 Results with the learning data	67
5.4.9 Results with unseen data	67
5.5 Comparison of the results	68
5.6 Conclusions	72
References	72

6 A practical application of neural modelling and predictive control   **74**
J.T. Evans, J.B. Gomm, D. Williams, P.J.G. Lisboa and Q.S. To

6.1 Abstract	74
6.2 Introduction	74
6.3 Data conditioning	76
6.4 The process	7176
6.5 Developing a neural network model of the system	79
6.5.1 Neural network topology	79
6.5.2 Network training	79
6.5.3 Model validation	81
6.5.4 Real process results	82
6.6 Neural predictive control: a preliminary study	84
6.6.1 On-line results	85
6.7 Conclusions and further work	87
Acknowledgements	88
References	88

7 A label-driven CMAC intelligent control strategy   **89**
A.J. Lawrence and C.J. Harris

7.1 Abstract	89
7.2 Introduction	89
7.3 The CMAC algorithm	90
7.4 Controller design and operation	92
7.4.1 Control scheme	92

7.4.2 The CMAC tuning module	94
7.5 Results	97
7.6 Conclusions and further work	99
References	102
8 Neural network controller for depth of anaesthesia	**104**
D.A. Linkens and H.U. Rehman	
8.1 Introduction	104
8.2 Background and history	105
8.3 Present work and discussion	105
8.4 Conclusions	115
References	115
Index	117

Preface

Interest in artificial neural networks began in the early 1940s when pioneers, such as McCulloch and Pitts and Hebb, investigated networks based on the neuron and attempted to formulate the adaptation laws which applied to such systems. During the 1950s and 1960s several basic architectures were developed and a background body of knowledge was buit up from many diverse disciplines: biology, psychology, physiology, mathematics and engineering. General interest in the subject waned after the analysis of the perceptron by Minsky and Papert highlighted the limitations of several of the models. However, several groups did continue and by the mid 1980s the work of Hopfield and of Rumelhart gave a renewed impetus to the area. Since then the number of papers published, conferences organized and journals devoted exclusively to neural network research has mushroomed.

Neural networks have several important characteristics which are of interest to control engineers:

- Modelling. Because of their ability to be trained using data records for the particular system of interest, the major problem of developing a realistic system model is obviated.
- Non-linear systems. The networks possess the ability to 'learn' non-linear relationships with limited prior knowledge about the process structure. This is possibly the area in which they show the greatest promise.
- Multivariable systems. Neural networks, by their very nature, have many inputs and many outputs and so can be readily applied to multivariable systems.
- Parallel structure. The structure of neural networks is highly parallel in nature. This is likely to give rise to three benefits: very fast parallel processing, fault tolerance and robustness.

The great promise held out by these unique features is the main reason for the enormous interest which is currently being shown in this field.

This book arises from a very successful colloquium that was held in Liverpool and was jointly organized by Liverpool John Moores University and the Merseyside branch of the Institute of Measurement and Control. The colloquium was attended by over 80 delegates from the UK, of whom approximately half were from industry. The book is of particular interest to those currently investigating neural networks and also to practising engineers, for whom the applications may reveal that neural networks could be a useful tool in their field. Also, the book may be suitable to support courses, particularly at postgraduate level, where the course emphasis is on implementation and applications of neural networks. Rapid progression of this field makes it difficult for the text to provide a comprehensive coverage of the subject. However, the book does provide a valuable insight into current reseach activity in artificial neural network applications to modelling and control.

The topics in this book can be split into three parts: (1) a general introduction to the field plus a theoretical discussion; (2) applications of neural networks in process modelling and estimation; and (3) applications of various neural network control strategies. Chapter 1 is for the reader who is either new to, or has limited knowledge of, the area. It gives a gentle introduction to the subject and overviews the configurations which are most applicable to modelling and control. The second chapter deals with the historical evolution of modern non-linear networks. The characteristics of these networks are discussed together with various issues regarding practical applications to estimation and control.

Approaches of applying neural networks to process modelling and estimation are described in Chapters 3, 4 and 5. The first of these deals with the effectiveness of recurrent networks for the identification of linear systems and it demonstrates that to model higher-order systems successfully the basic Elman net has to be modified. Chapter 4 presents a new paradigm for enhancing the training of feedforward networks. The training procedure is described and implications of the training philosophy are discussed. Some results of the technique applied to industrial data are presented to demonstrate the effectiveness of the technique. An interesting feature of the paradigm is that it incorporates a systematic procedure for determining the number of neurons in the hidden layer of the network. The following chapter by Tsaptsinos *et al.* deals with an estimation problem. It compares a conventional identification approach to the modelling of a fermentation process with a neural network technique.

The final three chapters present methodologies for achieving neural control. Evans *et al.* in Chapter 6 bridge the modelling and control aspects and describe a neural modelling procedure which leads to an actual working predictive controller. Chapter 7 introduces the cerebellar model articulation controller (CMAC). This belongs to a class of neural network which consists of a fixed non-linear input layer coupled to an adjustable linear output layer.

The particular CMAC-based adaptive control strategy described here is being developed as part of a project to produce an intelligent controller for aeronautical gas turbine engines. The chapter describes an adaptive control strategy and presents some preliminary results. The last chapter demonstrates the very wide potential applicabiity of the neural network approach by presenting a unique problem which has some very severe constraints. A back-propagation learning paradigm is described which has been developed to control depth of anaesthesia.

Finally, the editors would like to express their gratitude to all the contributing authors, to the staff at the Institute of Measurement and Control and to all at Chapman & Hall who have helped bring this project to completion.

G.F. Page
J.B. Gomm
D. Williams

Contributors

J.T. Evans
Control Systems Research Group
School of Electrical and Electronic
 Engineering
Liverpool John Moores University
Byrom Street
Liverpool, L3 3AF, UK

J.B. Gomm
Control Systems Research Group
School of Electrical and Electronic
 Engineering
Liverpool John Moores University
Byrom Street
Liverpool, L3 3AF, UK

C.J. Harris
Aeronautics and Astronautics
 Department
Southampton University
Highfield
Southampton, SO9 5NH, UK

N.A. Jalel
Industrial Control Centre
Faculty of Engineering and Science
University of Westminster
115 New Cavendish Street
London, W1M 8JS, UK

A.J. Lawrence
Aeronautics and Astronautics
 Department
Southampton University
Highfield
Southampton, SO9 5NH, UK

J.R. Leigh
Industrial Control Centre
Faculty of Engineering and Science
University of Westminster
115 New Cavendish Street
London, W1M 8JS, UK

D.A. Linkens
Department of Automatic Control
 and Systems Engineering
University of Sheffield
P.O. Box 600
Sheffield, S1 4DU, UK

P.J.G. Lisboa
Department of Electrical
 Engineering and Electronics
University of Liverpool
P.O. Box 147
Liverpool, L69 3BX, UK

Contributors

X. Liu
Intelligent Systems Research Laboratory
School of Electrical, Electronic and Systems Engineering
University of Wales College of Cardiff
P.O. Box 904
Cardiff, CF1 3YH, UK

G.F. Page
Control Systems Research Group
School of Electrical and Electronic Engineering
Liverpool John Moores University
Byrom Street
Liverpool, L3 3AF, UK

C. Peel
Department of Chemical and Process Engineering
University of Newcastle upon Tyne
Newcastle upon Tyne, NE1 7RU, UK

D.T. Pham
Intelligent Systems Research Laboratory
School of Electrical, Electronic and Systems Engineering
University of Wales College of Cardiff
P.O. Box 904
Cardiff, CF1 3YH, UK

H.U. Rehman
Department of Automatic Control and Systems Engineering
University of Sheffield
P.O. Box 600
Sheffield, S1 4DU, UK

M.T. Tham
Department of Chemical and Process Engineering
University of Newcastle upon Tyne
Newcastle upon Tyne, NE1 7RU, UK

Q.S. To
Department of Electrical Engineering and Electronics
University of Liverpool
P.O. Box 147
Liverpool, L69 3BX, UK

D. Tsaptsinos
Industrial Control Centre
Faculty of Engineering and Science
University of Westminster
115 New Cavendish Street
London, W1M 8JS, UK

D. Williams
Control Systems Research Group
School of Electrical and Electronic Engineering
Liverpool John Moores University
Byrom Street
Liverpool, L3 3AF, UK

M.J. Willis
Department of Chemical and Process Engineering
University of Newcastle upon Tyne
Newcastle upon Tyne, NE1 7RU, UK

Introduction to neural networks 1

J.B. Gomm, G.F. Page and D. Williams

1.1 ABSTRACT

This chapter is intended to provide a general introduction to neural networks for the reader who is either new to, or has limited knowledge in, this area. Concepts of neural networks are introduced, with a background to biological aspects, and their attributes are described. Many types of neural network exist and the configurations that are most applicable to the context of this book, modelling and control, are overviewed. These networks are the multi-layer perceptron, Hopfield network and Kohonen network.

1.2 INTRODUCTION

Artificial neural networks have emerged from studies of how human and animal brains perform operations. The human brain is made up of many millions of individual processing elements, called neurons, that are highly interconnected. A schematic diagram of a single biological neuron is shown in Fig. 1.1 Information from the outputs of other neurons, in the form of electrical pulses, are received by the cell at connections called synapses. The synapses connect to the cell inputs, or dendrites, and the single output of the neuron appears at the axon. An electrical pulse is sent down the axon (i.e. the neuron 'fires') when the total input stimuli from all of the dendrites exceeds a certain threshold [1, 2].

Artificial neural networks are made up of individual models of the biological neuron (artificial neurons or nodes) that are connected together to form a network. The neuron models that are used are typically much simplified versions of the actions of a real neuron. Information is stored in

2 Introduction to neural networks

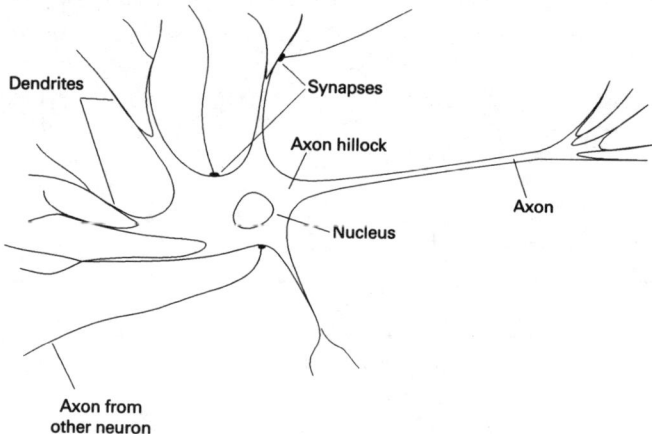

Fig. 1.1 Biological neuron.

the network often in the form of different connection strengths, or weights, associated with the synapses in the artificial neuron models.

Among the numerous attributes of neural networks that have been found in many application areas are [3]:

- inherent parallelism in the network architecture due to the repeated use of simple neuron processing elements. This leads to the possibility of very fast hardware implementations of neural networks.
- capability of 'learning' information by example. The learning mechanism is often achieved by appropriate adjustment of the weights in the synapses of the artificial neuron models.
- ability to generalize to new inputs (i.e. a trained network is capable of providing 'sensible' outputs when presented with input data that has not been used before).
- robustness to noisy data that occurs in real world applications.
- fault tolerance. In general, network performance does not significantly degenerate if some of the network connections become faulty.

Conventional programming techniques are significantly better than humans at performing tasks requiring a high degree of numerical computation and repeatable steps that can be accurately pre-specified. However, humans still far exceed the performance of these methods in applications that are poorly defined, either because the problem is extremely complex or simply that exact solution rules are not known (e.g. speech and image recognition, plant monitoring and fault diagnosis). The above attributes of neural networks indicate their potential in solving these problems. Hence, the considerable interest in neural networks that has occurred in recent years is not only due to significant advances in computer processing power that has enabled their implementation, but also because of the diverse possibility of application areas.

Artificial neuron model 3

Many different types of neural network are available and only the architectures that are most applicable to the context of this book, modelling and control, are described. The basic building block of these networks, the artificial neuron model, is first introduced. This is followed by overviews of three networks architectures: the multi-layer perceptron, Hopfield and Kohonen networks. General descriptions of training algorithms for these networks are also given.

1.3 ARTIFICIAL NEURON MODEL

The most commonly used neuron model is depicted in Fig. 1.2 and is based on the model proposed by McCulloch and Pitts in 1943 [4]. Each neuron input, $x_1 - x_N$, is weighted by the values $w_1 - w_N$. A bias, or offset, in the node is characterized by an additional constant input of 1 weighted by the value w_0. The output, y, is obtained by summing the weighted inputs to the neuron and passing the result through a non-linear activation function, $f(\ )$:

$$y = f\left(\sum_{i=1}^{N} w_i x_i + w_0\right). \tag{1.1}$$

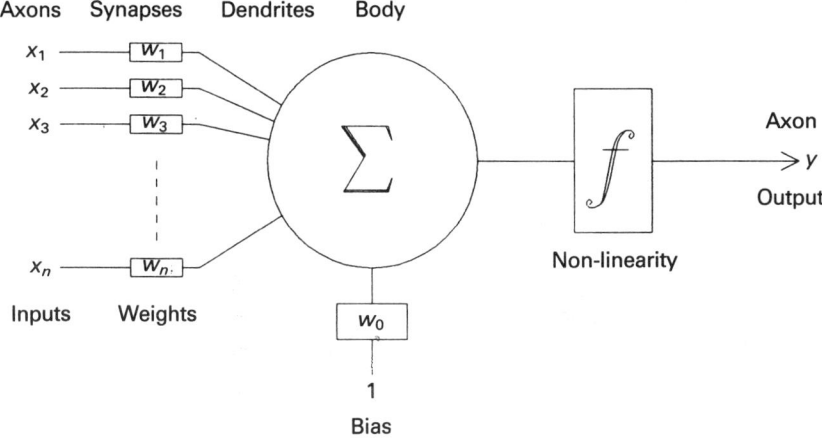

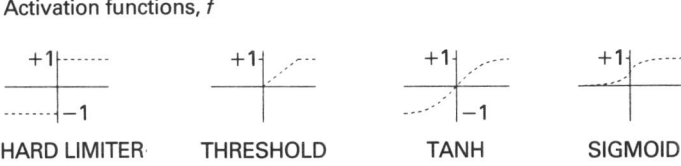

Fig. 1.2 McCulloch–Pitts neuron model.

Various types of non-linearity are possible and some of these are shown in the diagram (e.g. hard limiter, threshold logic, sigmoidal and tanh functions).

1.4 MULTI-LAYER PERCEPTRON

The most popular neural network architecture is the multi-layer perceptron (MLP). The network consists of an input layer, a number of hidden layers (typically only one or two hidden layers are used) and an output layer as shown in Fig. 1.3. The output and hidden layers are made up of a number of nodes as described in section 1.3. However the input layer is essentially a direct link to the inputs of the first hidden layer and is included by convention. Sigmoidal activation functions for the nodes in the hidden and output layers are the most common choice, although variants on this are also possible. The outputs of each node in a layer are connected to the inputs of all of the nodes in the subsequent layer. Data flows through the network in one direction only, from input to output; hence, this type of network is called a feedforward network.

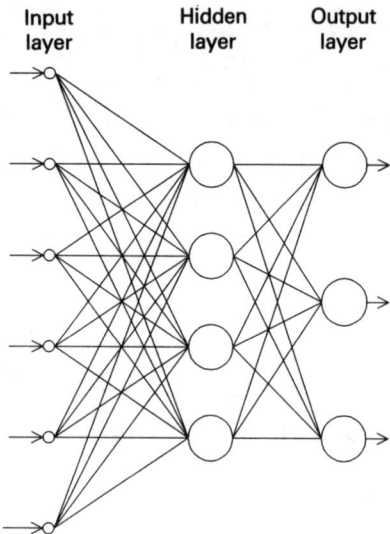

Fig. 1.3 Multi-layer perceptron with one hidden layer.

The network is trained in a supervised fashion. This means that during training both the network inputs and required, or target, outputs are used. A number of algorithms have been proposed for training the MLP and the most popular is the back-propagation algorithm [5, 6]. Briefly, with this algorithm a set of input and corresponding output data is collected that the network is required to learn. An input pattern is applied to the network and an output is generated. This output is compared to the corresponding target output and an error is produced. The error is then propagated back through the network, from output to input, and the network weights are adjusted in such a way as to minimize a cost function, typically the sum of the errors squared. The procedure is repeated through all the data in the training set and numerous passes of the complete training data set are usually necessary before the cost function is reduced to a sufficient value.

An important feature of the MLP is that this network can accurately represent any continuous non-linear function relating the inputs and outputs [7, 8]. Hence, the MLP network exhibits potential for many applications, including modelling and control of real non-linear processes.

1.5 HOPFIELD NETWORK

The Hopfield network consists of two layers, an input layer and a Hopfield layer (Fig. 1.4). Each node in the input layer is directly connected to only one node in the Hopfield layer. The nodes in the latter layer are neuron models previously described (section 1.3) with either hard limiting [9] or sigmoidal activation functions [10]. The outputs of these nodes are weighted and fed back to the inputs of all of the other nodes.

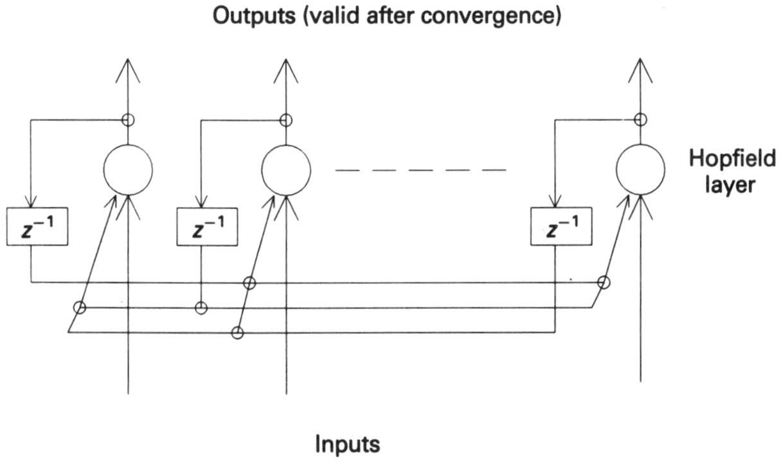

Fig. 1.4 Hopfield network. z^{-1} is the unit delay operator.

Operation of the Hopfield network is as follows. During training, the network output is often required to be the same as the input. Connection strengths are weakened, by reducing the corresponding weight values, if the output of a neuron is different from the input, and strengthened when the converse is true. The trained network is used by applying an input pattern to the network. The network outputs are then continually fed back through the weights until a convergence criterion is met, typically when there are no changes at the network output nodes on successive iterations. This is the final network output for the input pattern.

Binary input and output values, often represented as $+1$ and -1, are usually used with the Hopfield network. Applications have therefore been found where binary data is frequently present, such as in image processing for reconstructing images from noisy data and for pattern recognition in images irrespective of size, orientation and position [11]. The presence of feedback in this network also makes it of interest for application to modelling and control of dynamical systems [12, 13].

1.6 KOHONEN NETWORK

The main distinguishing feature of this network, from the MLP and Hopfield networks, is that no output data is required for training. A Kohonen network is constructed of a fully interconnected array of neurons (i.e. the output of each neuron is an input to all neurons, including itself) and each neuron receives the input pattern (Fig. 1.5). There are two sets of weights: an adaptable set, to compute the weighted sum of the external inputs, and a fixed set between neurons, that controls neuron interactions in the network.

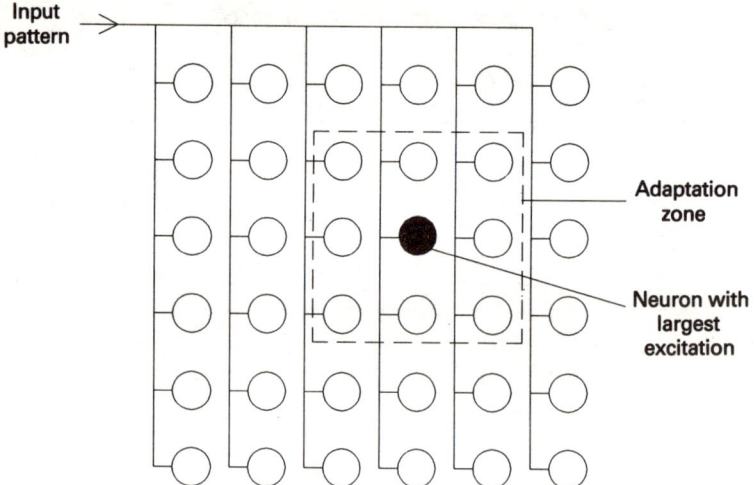

Fig. 1.5 Kohonen network.

Training involves applying an input pattern to the network, consisting of a set of continuous-valued data, and the output of each neuron is computed. The neurons are then allowed to interact with each other and the neuron that responds most to the input stimuli (i.e. the one that has the largest output) is found. Only this neuron and neighbourhood neurons, within a certain distance, are allowed to adjust their weights to become more responsive to the particular input. This form of training has the effect of organizing the 'map' of the output nodes such that different areas of the map will respond to different input patterns. Hence, the Kohonen network has self-organizing properties and is capable of recognition. Application examples of the Kohonen network include recognition of images and speech signals [1, 14].

1.7 CONCLUSIONS

Artificial neural networks, although stemming from biological studies of the brain, are currently far from accurately emulating the performance of the human brain. However, recent research has shown that they do exhibit useful properties and have much potential for solving problems in which more conventional programming techniques have met with great difficulty. Applications of artificial neural networks in areas such as image processing and pattern recognition are now fairly well founded. In process modelling and control, systems are often highly non-linear, complex and poorly characterized. Therefore, the features of neural networks also make them an attractive tool for applications in this area and it is emerging that they will become established in certain aspects of this field.

REFERENCES

1. Allinson, N.M. (1990) Neurons, N-tuples and faces. *Comp. Contr. Engng. J.*, **July**, 173–83.
2. Melsa, P.J.W. (1989) Neural networks: a conceptual overview. Internal Report: Tellabs Research Center, USA, August, Report No. TRC-89-08.
3. Lippmann, R.P. (1987) An introduction to computing with neural nets. *IEEE ASSP Magazine*, **April**, 4–22.
4. McCulloch, W.S. and Pitts, W.H. (1943) A logical calculus of the ideas immanent in nervous activity. *Bull. Math. Biophy.*, **5**, 115–33.
5. Werbos, P.J. (1974) Beyond regression: new tools for prediction and analysis in the behavioral sciences. Ph.D. Thesis in Applied Mathematics, Harvard University.
6. Rumelhart, D.E., Hinton, G.E. and Williams, R.J. (1986) Learning representations by back-propagating errors. *Nature*, **323**, 533–6.
7. Cybenko, G. (1989) Approximations by superpositions of a sigmoidal function. *Mathematics of Control, Signals and Systems*, **2**, 303–14.
8. Funahashi, K. (1989) On the approximate realization of continuous mappings by neural networks. *Neural Networks*, **2**, 183–92.
9. Hopfield, J.J. (1982) Neural networks and physical systems with emergent collective computational abilities. *Proc. Natl. Acad. Sci. USA*, **79**, 2554–8.
10. Hopfield, J.J. (1984) Neurons with graded response have collective computational properties like those of two-state neurons. *Proc. Natl. Acad. Sci. USA*, **81**, 3088–92.

11. Perantonis, S.J. (1992) Higher-order neural networks for invariant pattern recognition. In *Neural Networks: Current Applications* (ed. P.J.G. Lisboa), Chapman & Hall, London, Chap. 11, pp. 197–231.
12. Narendra, K.S. and Parthasarathy, K. (1990) Identification and control of dynamical systems using neural networks. *IEEE Trans. Neural Networks*, **1**(1), 4–27.
13. Zbikowski, R. and Gawthrop, P.J. (1992) A survey of neural networks for control, in *Neural Networks for Control and Systems* (ed. K. Warwick, G.W. Irwin and K.J. Hunt). Peter Peregrinus Ltd., London, Chap. 3, pp. 31–50.
14. Kohonen, T. (1988) *Self-organization and Associative Memory*, 2nd edn, Springer-Verlag, Berlin and Heidelberg.

Neural networks for control: evolution, revolution or renaissance

2

P.J.G. Lisboa

2.1 ABSTRACT

The last six years have experienced a revival of numerical algorithms under the general banner of artificial neural networks. Many of these methods can be regarded as extensions of the NARMAX approach, but they also possess associative properties which may increase their usefulness. The historical evolution of modern non-linear networks is reviewed first. The novel aspects of the approach are then discussed, together with a brief overview of available control strategies. Finally, a number of issues are raised regarding the practical application of this approach to estimation and control.

2.2 INTRODUCTION

The field of control spans a wide range of applications. At one extreme there is scheduling of regulators to control electromechanical movement in manufacturing processes. These tasks may be tackled using optimization methods through gradual adjustments of steady state conditions. At the other extreme is the requirement for precision control of very fast processes, as in the aeronautics industry. These processes are usually accurately modelled using one or more linearized equations, and are amenable to solutions using finely tuned controllers of very high order. The remainder of this chapter is concerned with the middle range of this spectrum, comprising the control of dynamical systems in manufacturing and the process industries, and in other areas including medical applications. This is the realm of plant that is typically inaccurately modelled, time varying, and subject to disturbances, but with moderately slow response times in relation to modern computer processing speeds. The niche of interest to artificial neural networks comprises those processes which involve also a degree of non-linearity, which is the main limitation of the otherwise successful conventional control methods, especially where the non-linearity is of an unknown structure, or is very severe.

2.3 HISTORICAL BACKGROUND

The main distinguishing feature of artificial neural networks, and arguably one of the few characteristics shared by the vast majority of paradigms classed under this heading, is that they are structured as non-linear adaptive filters, comprising layers of non-linear adaptive processing elements. This is described symbolically in Fig. 2.1. The original perceptron [1], which is the role model of modern network achitectures, had just two layers of nodes. The first layer was called the sensory layer, because it simply relayed the

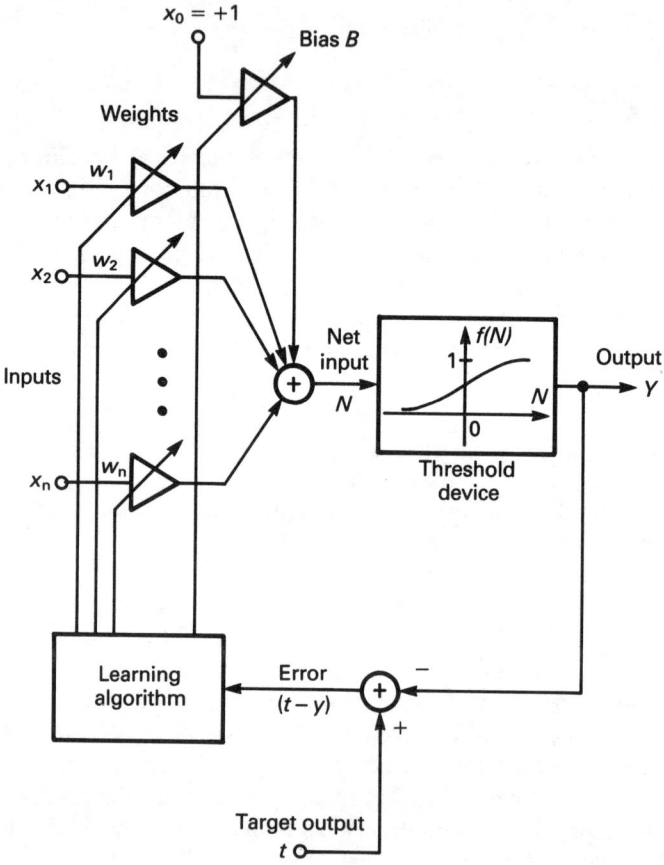

Fig. 2.1 Representation of a neural network as an adaptable filter.

Historical background 11

inputs received from an array of sensors, usually focused on an image, to the next layer. The second layer of nodes was termed the associative layer, because it formed associations of patterns of excitation at the sensory layer. Once the precise role of the associative nodes was elucidated [2], in a way which today is thought of in terms of linear discriminants formed around clusters of patterns belonging to each class [3], it became clear that the network as a whole acted as a matrix multiplier and was therefore unable to distinguish between non-linearly separable input patterns. A celebrated example of this was the XOR problem, although it was realized at the time that this function could be realized using networks with a single associative layer of nodes, by extending the dimensionality of the input space using x, y and also the product of x and y. However, this was regarded as generally impracticable for real applications, which in fact turns out not to be the case.

This leads us to discuss the three basic elements from which evolved the modern family of neural computing methods for dynamical systems. First, there is the 'adaptive transversal filter' introduced in 1960 by Widrow and Hoff [4], and shown schematically in Fig. 2.2. This is a linear filter consisting of a tapped delay line, commonly used today to enter data into neural network estimators for dynamical systems. The description of the signal as a time series is familiar to practitioners in modern digital signal processing. This filter architecture was the first to develop the concept of layers of nodes, linked by scalar factors, and trained by gradient descent according to a quadratic error function defined at the output nodes of the network. The filter acts like an estimator by retrieving the nominated target response to each input pattern, which shows a window comprising the recent history of the signal.

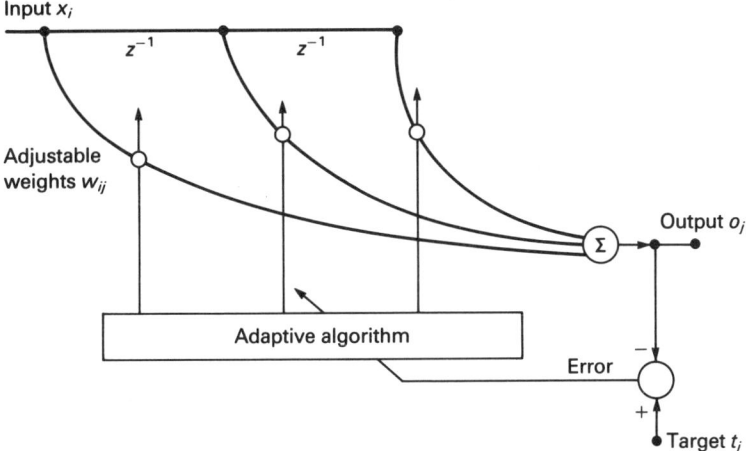

Fig. 2.2 Linear adaptive filter using a tapped delay line.

12 Neural networks for control

The limitation of this approach was that an appropriate description of non-linear functions requires an additional layer of processing elements. The difficulty in extending this concept to filters with multiple layers was perceived at the time as the absence of an explicit error function for the nodes in the intermediate layers, hidden away from direct contact with either the inputs or target outputs. This was known as the credit assignment problem [2] and remained a major obstacle to further development of this approach. The historical development turned to more sophisticated methods for estimating the coefficients of linearized models, using auto-regressive and predictive methods.

An independent process of evolution took place regarding associativity. Here the emphasis was on adequate scheduling of non-linear processes. This was achieved using a generalized look-up table, where a vector representing the current plant state is multiplied by a matrix of correlations [5] between the plant states and the corresponding desired control actions. The matrix operates as a look-up table storing the desired response for a set of nominated plant states, but in addition it serves the purpose of interpolating the control actions when the plant state is a mixture of several of the nominated ones. This interpolation represents the interpretation of associative behaviour in terms of a function map between the input state and the space of control actions.

Matrix associative memories [6–8] mark the emergence of the pattern recognition approach to artificial intelligence, and control in particular. This approach was inspired by biological considerations [8], and has found applications in control [9], acting like a variable-structure control schedule. The matrix is built up by correlation training, which is a process closely related to gradient descent, as follows. The elements in the matrix are the summations of the outer product of the input and output vectors representing the desired relationships. However, this is just the starting step in iterative gradient descent, since:

$$\delta w_{ij} = -\frac{\eta}{2}\frac{d}{dw_{ij}}\left(t_j - \sum_j w_{ij} x_i\right)^2$$

$$= \eta \left(t_j - \sum_j w_{ij} x_i\right) x_i$$

$$= t_j x_i |_{w_{ij}=0, \eta=1}. \qquad (2.1)$$

The property of association is, therefore, equally prevalent in networks trained by gradient descent.

Another method inspired by learning processes in living mechanisms is reinforcement learning [10]. This is the idea that biological conditioning arises from the strengthening of links between processors by continual reinforcement of the prediction of an eventual outcome. The prediction that a pole will fall out of balance after T time steps, for instance, can be 'bootstrapped' by considering the difference between successive predictions at time t and $t+1$, in addition to the actual reinforcement arising from the eventual fall of the pole [11], multiplied by a forgetting factor. The estimate of the pole falling in one, two, three, etc. time steps from now is reflected in the following change in weights.

$$E = [t(t) - o(t)]^2 + \sum_n \lambda^n [o(t) - o(t-1)]^2$$
$$\approx (1-\lambda)[t(t) - o(t)]^2 + \lambda[o(t) - o(t-1)]^2. \quad (2.2)$$

When the forgetting factor is zero, the usual gradient descent methods result, but the minimization of this error function is often achieved for a finite value. This is a standard result from numerical optimization, and applies equally well to minimizing errors in plant predictors, especially concerning optimal generalization of estimates for test data not used during training. Its importance here lies in the relevance of these methods in the development of controllers of processes where the success or otherwise of the control action is determined qualitatively, so that once the network begins to train, error correcting signals are scarce, forcing the system to self-learn [12].

Finally, an early precursor of artificial intelligent control was the BOXES method of Michie and Chambers [13]. The importance here was the quantization of the input parameter space into cells, so that the plant state at any time consisted of a binary vector with only one non-zero component. The period between making a control decision at a particular cell, and an eventual control failure, was recorded and the decision taken on subsequent visits to that cell was that which measured the furthest from the failure point. An associative version of this approach to decision-making has also been achieved using hashing methods [14]. This led to the cerebellar model articulation controller (CMAC) by analogy with associative processes in the cerebellum [15].

2.4 ESTIMATION

The revolutionary aspect of the current revival owes more to the low cost of computer power, which makes simulation of complex non-linear systems possible, than it does to any one conceptual leap. A possible exception to this was the solution of the credit assignment problem by old-fashioned gradient descent, when it was realized that this led to an effective error measure at the output of every node in the network [16], including the hidden layers, and in fact even further back to the input nodes, if necessary, to achieve desired target outputs. The error at a hidden node accumulates the error present at the network output via the strength of its links to those nodes, hence the term back-error propagation (BEP) [17].

14 Neural networks for control

This development opened the doors to a general-purpose algorithm probably capable of modelling non-linear processes [18], even though this approach is not alone since the previously mentioned CMAC model is equally capable of doing that [19]. In addition, the associative properties of both sets of non-linear adaptive filters are expected to result in robust and effective performance in both cases, which it is hoped will supersede conventional methods of non-linear estimation and control, particularly in processes where the non-linearities are hard to model accurately, and combine with noise and time variability.

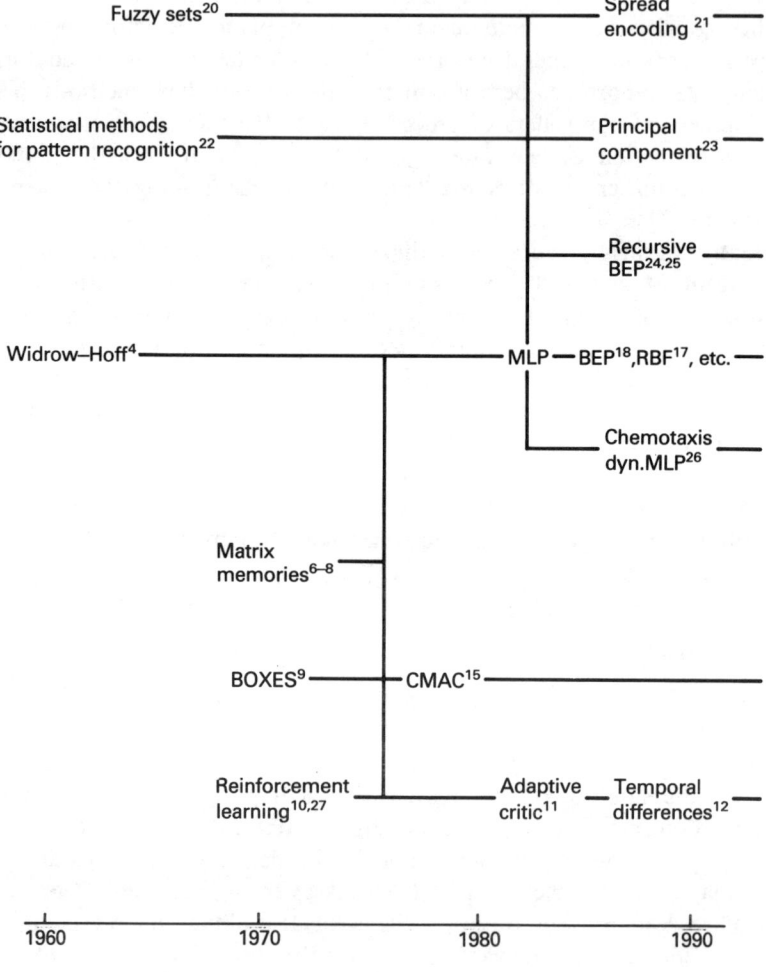

Fig. 2.3 Genealogy of neural networks for control.

A genealogy of neural networks is shown in Fig. 2.3. Among these, the multi-layered perceptron network (MLP) is by far the most commonly used neural computing paradigm, largely because it is versatile and easy to implement. The form of the non-linear functions which replace the simple summation in Fig. 2.1 varies, including the use of radial basis functions (RBF) [18] in combination with linear output nodes. For a review of several of these algorithms see reference [28].

A distinction must be drawn here between one-step predictors, for which straightforward back-error propagation applies, and recurrent MLPs, which require different training algorithms [24, 25]. The difference between these processes is indicated in Fig. 2.4, and the estimation process is represented symbolically in Fig. 2.5.

The usefulness of a plant estimator in predictive control depends strongly upon its ability to model the plant accurately over several time steps, even if a receding horizon is used, so that only the first of each sequence of control actions is acted upon [26]. A method of training a conventional MLP in

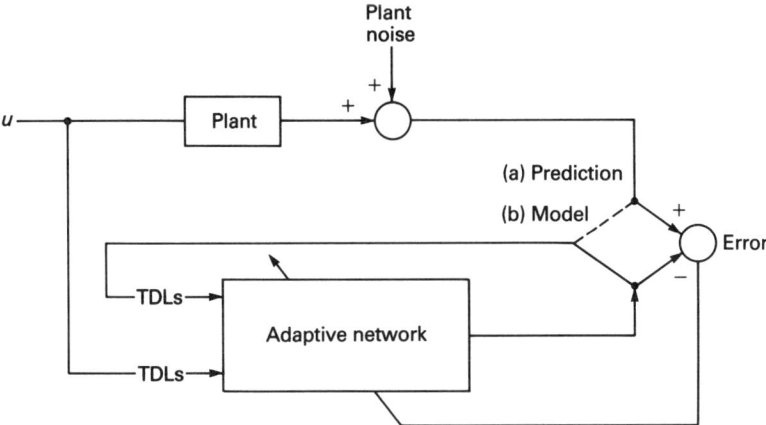

Fig. 2.4 Plant estimators: (a) one-step-ahead predictor, (b) recurrent MLP.

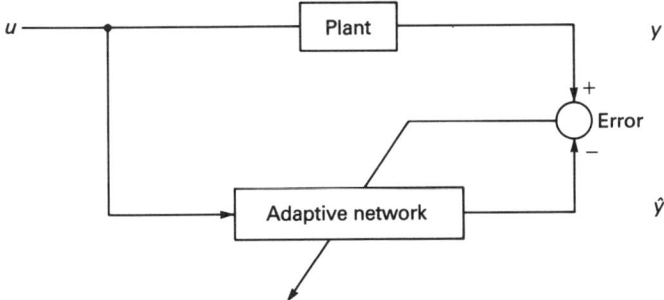

Fig. 2.5 Symbolic representation of a neural plant estimator.

such a way that the activity of the network when testing its response to a new control sequence is self-sustained, thereby acting as a genuine model of the plant, is to increase the accuracy achieved during training, by using spread encoding [21, 29].

The problem of accuracy of representation becomes crucial when recursive networks are considered, since the activity of individual units at the output layer of the network must return to the corresponding input units, with the same dynamic range, which is usually between zero and unity. The improvement in the accuracy of this representation for analogue variables is achieved by extending the dimensionality of the neural network, in a way that is reminiscent of fuzzy-logic membership functions [30]. This form of coding is illustrated in Fig. 2.6, where it is compared to the use of non-linear discriminant surfaces by adding $x^{1/2}$ and x^2 to the linear term in x at the input and output layers of the network.

A variable x, with an arbitrary finite range $[x_{min}, x_{max}]$ is mapped onto a sliding pattern of activation of N input nodes, with additional nodes either side to contain any spill-over resulting from the use of a mapping function with wide support. This creates a discrete distribution, which can be made to represent the mean value of the continuous distribution within each class interval, rather than the mean probability density in that interval as is usually the case. This is achieved by defining the excitation of each node as:

$$\psi_i = \frac{\int_{a_i-1/2}^{a_i+1/2} [af(a)] \, da}{a_i} \qquad (2.3)$$

satisfying the requirement that:

$$\sum_i a_i \psi_i = \int [af(a)] \, da = \langle a \rangle. \qquad (2.4)$$

However, the resulting values of ψ_i no longer add up to unity. Therefore they can no longer be interpreted as forming a probability distribution. The activation of each input node i can be evaluated by integration by parts:

$$a_i \psi_i = [aF(a)]_{a_i-1/2}^{a_i+1/2} - \int_{a_i-1/2}^{a_i+1/2} F(a) \, da \qquad (2.5)$$

where $F'(a) = f(a)$.

The advantages of spread encoding over the use of a different node for each variable include noise reduction by the central limit theorem in the reconstruction of the output signal, as well as a confidence measure defined by the width of the output prediction patterns. An application of spread encoding to estimation of the dynamics of a non-linear process is described in detail in Chapter 6 of this book.

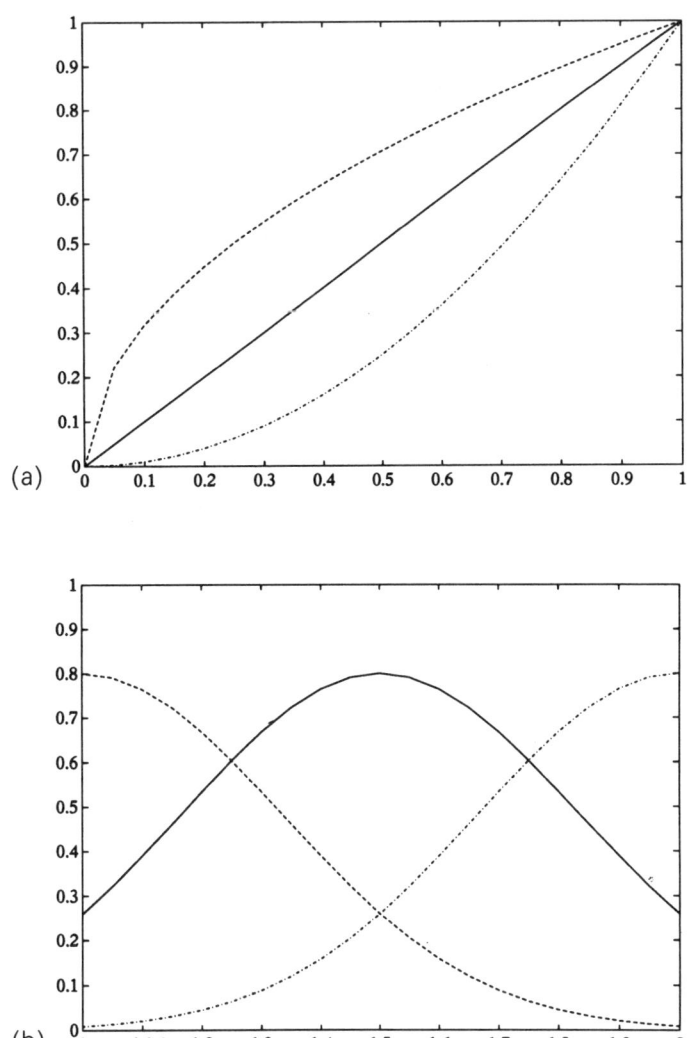

Fig. 2.6 Data representation for MLPs: (a) non-linear discriminants, (b) spread encoding.

In addition, the associative partition of the input space and the increase in its dimensionality may also improve the discriminating ability of the network. A more extreme version of this is accomplished in CMAC to fit non-linear maps using a single layer of adjustable weights (Fig. 2.7). After first extending the dimensions of the input space at the level of the sensory layer, hashing mechanisms have long been known to be an efficient mechanism for reducing phase space back to a manageable size while creating useful associative properties [14].

18 Neural networks for control

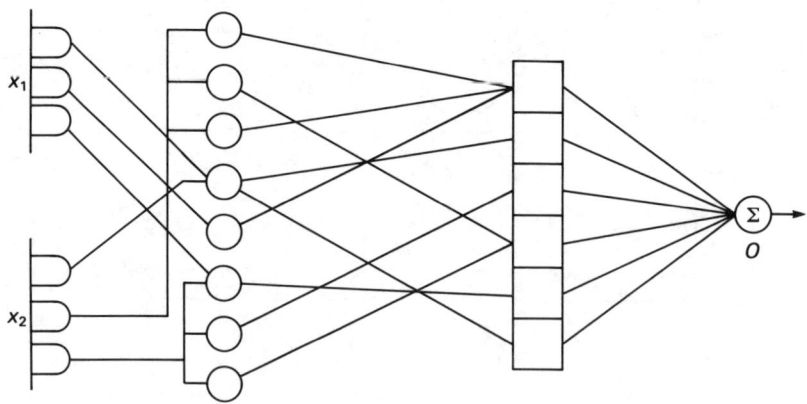

Fig. 2.7 CMAC structure.

An alternative method for improving the training accuracy and speed is to reorganize the input data using standard statistical decorrelation techniques [23]. Still, the problem remains of interpreting the model that is obtained.

A solution to this problem is to build the network dynamics into the links themselves, by replacing the proportional link gains with elemental blocks:

$$w_{ij}\frac{e^{-s\Delta T_{ij}}}{1+s\tau_{ij}}. \tag{2.6}$$

Computation of these coefficients is not possible by gradient descent, and alternative methods must be used [26]. Formal expansions of arbitrary non-linear systems in terms of basis functions generated by MLPs have also been discussed in the literature [31].

2.5 CONTROL

The availability of practical general-purpose adaptive filters for non-linear process estimation makes it possible to implement a variety of control structures, which herald a renaissance of earlier schemes, now released from the constraints of piecewise linearity. In this section, the main themes running through the neural computing approach to adaptive and non-linear process control are briefly reviewed.

Control 19

2.5.1 ADAPTIVE CRITIC (FIG. 2.8)

This is a general-purpose control method aimed at minimizing long-range prediction errors. It employs a plant state evaluator which predicts the likelihood of future plant feedback signals [11, 12]. This type of control has provable convergence properties [12], and it is suitable for controlling plant with qualitative or scarce feedback information.

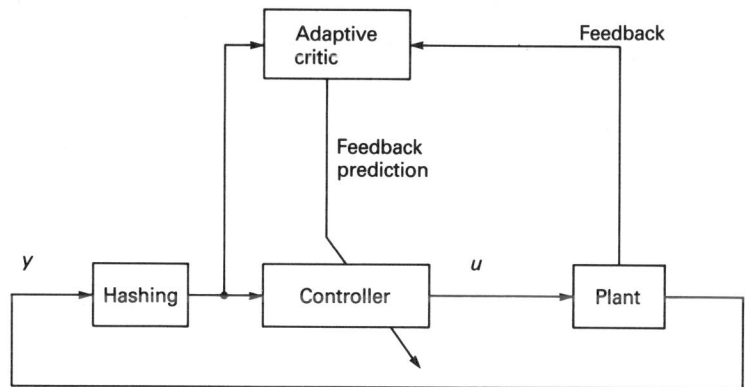

Fig. 2.8 Adaptive critic for eventual feedback.

2.5.2 COPY CONTROL (FIG. 2.9)

There are situations where it is advantageous to copy an existing controller, for instance where a human controller may be the most effective, yet the controlling task may be laborious or repetitive and therefore best suited to automatic implementation, leaving the operator free for other more

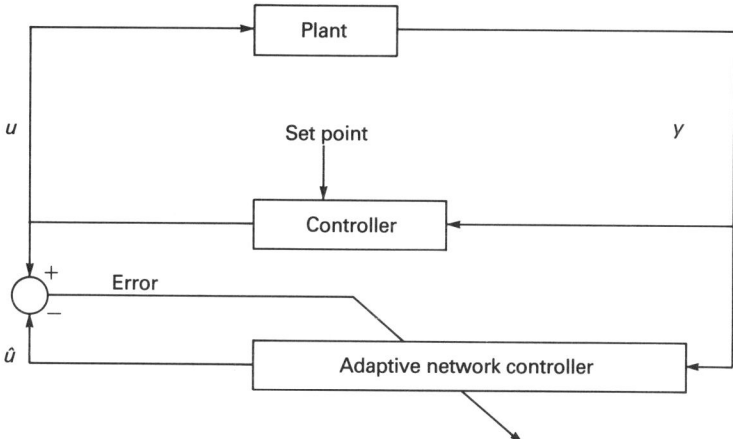

Fig. 2.9 Copy control strategy.

important tasks. Neural networks are then used mainly as associative memories, storing a previous history of controller commands, and interpolating between the sampled plant states used during training.

2.5.3 PREDICTIVE CONTROL (FIG. 2.10)

The next step in controller design for dynamical systems is to estimate plant behaviour, and use these estimates to design an optimal controller [26], minimizing a cost function of the form:

$$J = \sum_t [\text{setpoint}(t) - y(t)]^2 + \mu[u(t) - u(t-1)]^2 \qquad (2.7)$$

This control structure does not require inversion of the model, and is therefore suited to control highly non-linear plant, as occurs, for example, in the chemical industries.

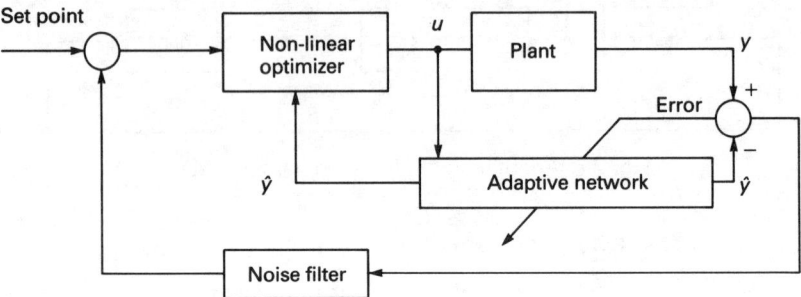

Fig. 2.10 Neural predictive control strategy.

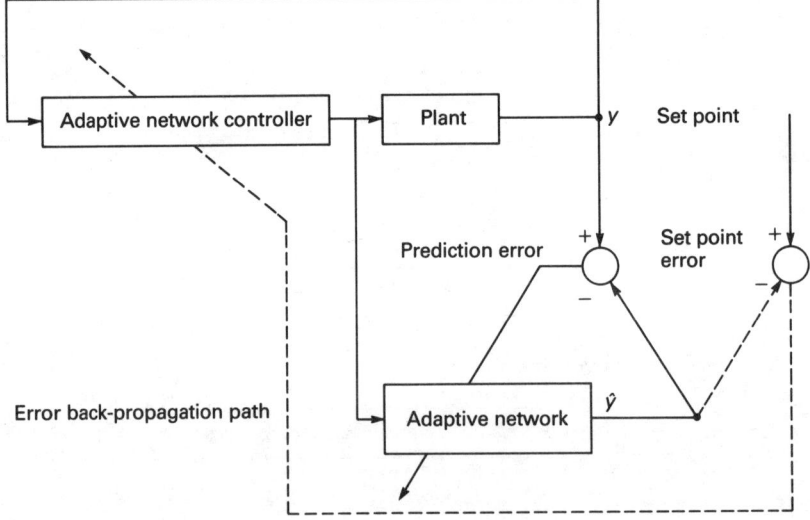

Fig. 2.11 Differentiating the model.

2.5.4 DIFFERENTIATING THE MODEL (FIG. 2.11)

This is a way of using the predictor adaptive network to compute the effective Jacobian of the plant, allowing the error in achieving the set point to propagate, through the plant, to correct the controller settings [32].

2.5.5 MODEL REFERENCE CONTROL (FIG. 2.12)

The block diagram in Fig. 2.12 shows the trimmings described in a paper [33] that preceded the publication of the BEP algorithm. At the time the

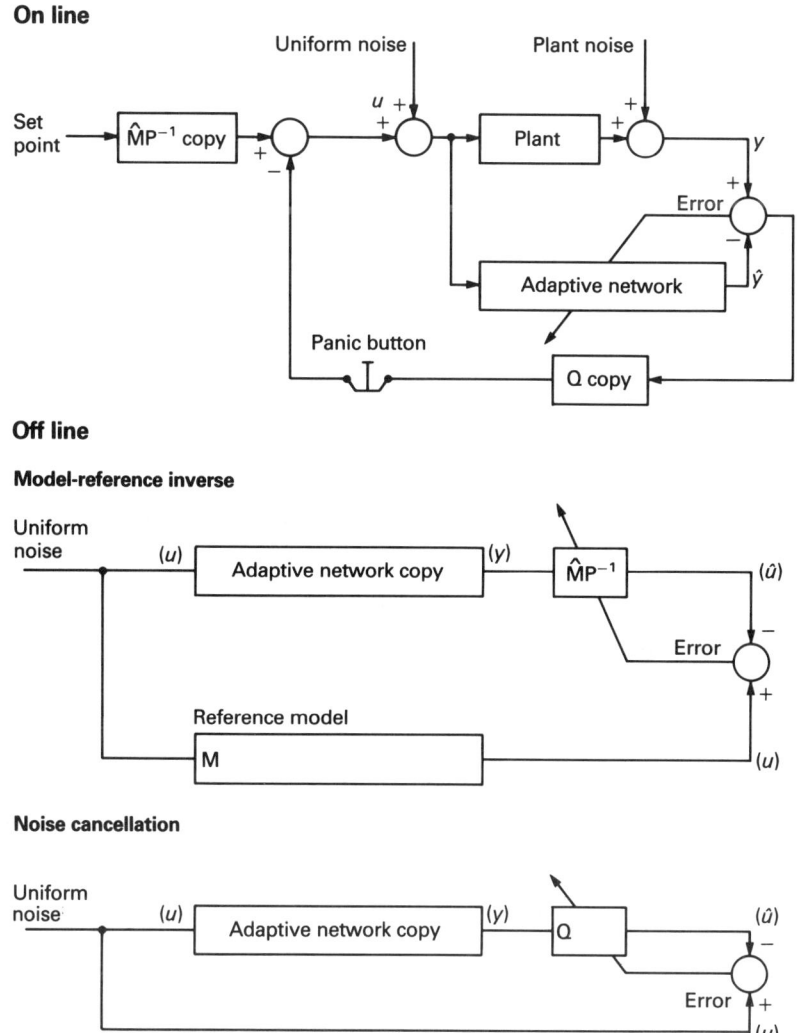

Fig. 2.12 Model reference control through plant inverse with noise cancellation.

approach was intended for adaptive control by linearizing the plant, but it can be extended using non-linear filters. A plant estimator is trained on line, for the purpose of inverting the plant. When the plant is not invertible, for instance non-minimum phase, this requires coupling it with a reference model, to regularize the joint transfer function using delays and lags in the reference model. The inverse is computed off line and updated at regular intervals. An additional plant inverse is often also required to counteract plant noise. Because of the nature of the process, this may only be well defined locally, i.e. for a narrow operating range, and may become unstable when the regime of operation suddenly changes. The original control strategy [33] already foresaw a manually operated panic button.

2.6 CONCLUSIONS

Non-linear adaptive filters are conveniently implemented using artificial neural networks. After evolving from generalized linear filters by combining them with associative networks and related models, the revolution in computer power has led to a renaissance of old methods now made feasible for practical applications. There are also some additional concepts, such as back-error propagation. In this particular case, its usefulness has been demonstrated mostly by experimentation since theoretical expectations might founder on a profusion of local minima, not to mention the very real problem of overfitting data by overtraining, which was noted in earlier studies of pattern recognition [34].

However, if neural computing is to mature into a fully practical approach to estimation and control, a number of issues need to be resolved.

In estimation, there is the practical task of ensuring that the network is trained to its optimal convergence point, beyond which its performance on blind tests, i.e. its generalization ability, begins to deteriorate. Additional theoretical questions remain about qualifying the precise range of operating conditions where the resulting model is valid, how to quantify the ability of the network design and training process to accurately represent the process dynamics, e.g. in terms of data sampling rates, the nature of the training signal to be used, and the nature of the plant non-linearity, plus all of the usual problems related to adaptive systems updating on line, where overtraining may occur at a particular operating condition, with disastrous consequences when this suddenly changes.

Regarding control, issues related to plant inversion and the choice of reference model are as pertinent here as in conventional adaptive control, only less clearly defined. There is in addition the problem of training speed, which afflicts some choices of training methods more than others, and may sometimes be exchanged for a reduction in the information storage capacity of the network by altering its size. Integration with other methods, conventional, fuzzy-logic or rule-based, are all possible and perhaps even desirable, since all of these methods perform different specialized functions.

Given the available array of neural computing tools, it is likely that further progress will be made in terms of experimentation on existing and new applications, complementing theoretical studies, currently in their infancy, of network dynamics.

ACKNOWLEDGEMENTS

The author wishes to acknowledge stimulating and informative discussions with the control systems research group at the Liverpool John Moores University, and also Q.S. To for carrying out the computer simulations presented here.

REFERENCES

1. Block, H.D. (1962) The perceptron: a model for brain functioning. *Rev. Mod. Phys.*, **34**, 123–35.
2. Minsky, M. and Papert, S. (1969) *Perceptrons*. MIT Press, Cambridge, Mass.
3. Lippmann, R.P. (1987) An introduction to computing with neural nets. *IEEE ASSP Mag.*, **April**, 4–22.
4. Widrow, B. and Hoff, M.E. (1960) Adaptive switching circuits. *Proc. IRE Western Electronic Show and Conv.*, **4**, 96–104.
5. Hebb, D.O. (1949) *The Organization of Behaviour*. Wiley, New York.
6. Nakano, K. (1972) Associatron – a model of associative memory. *IEEE Trans. Syst. Man and Cybern.*, **2(3)**.
7. Nakano, K. (1971) Learning process in a model of associative memory. In *Pattern Recognition and Machine Learning*, (ed. K.S. Fu), Plenum Press, New York, pp. 172–86.
8. Kohonen, T. (1989) *Self-Organization and Associative Memories*, 3rd edn. Springer-Verlag, New York.
9. Matsuoka, F. (1988) Holonics controller for air conditioners. Mitsubishi Electric ADVANCE, 27–29 March.
10. Mendel, J.M. and McLaren, R.W. (1970) Reinforcement learning control and pattern recognition systems. In *Adaptive, Learning, and Pattern Recognition Systems: Theory and Applications*, (ed. J.M. Mendel and K.S. Fu), Academic Press, New York, pp. 287–318.
11. Barto, A.G., Sutton, R.S. and Anderson, C.W. (1983) Neuronlike adaptive elements that can solve difficult learning control problems. *IEEE Trans. Syst. Man and Cybern.*, **13(5)**, 834–46.
12. Sutton, R.S. (1988) Learning to predict by the methods of temporal differences. *Mach. Learn.*, **3**, 9–44.
13. Michie, D. and Chambers, R.A. (1968) BOXES: An experiment in adaptive control. In *Machine Intelligence*, (ed. E. Dale and D. Michie), Oliver and Boyd, Edinburgh, pp. 137–52.
14. Andreae, J.H. (1977) *Thinking with the Teachable Machine*. Academic Press, London.

15. Albus, J.S. (1975) A new approach to manipulator control: The cerebellar model articulation controller (CMAC). *Trans. ASME, J. Dynamic Syst. Meas. Contr.*, **97**, 220–7.
16. le Cun, Y. (1985) A learning procedure for asymmetric threshold networks. *Proc. Cognitiva*, **85**, 599–604.
17. Rumelhart, D.E., Hinton, G.E. and Williams, R.J. (1986) Learning representations by back-propagating errors. *Nature*, **323**, 533–6.
18. Poggio, T. and Girosi, F. (1990) Networks for approximation and learning. *Proc. IEEE*, **78**(9), 1481–97.
19. Wong, Y.-F. and Sideris, A. (1992) Learning convergence in the Cerebellar Model Articulation Controller. *IEEE Trans. Neural Networks*, **3**(1), 115–21.
20. Zadeh, L.A. (1965) Fuzzy sets. *Inform. Contr.*, **8**, 338–53.
21. Lin, C.-T. and Lee, C.S.G. (1991) Neural-network-based fuzzy logic control and decision system. *IEEE Trans. Computers*, **40**(12), 1320–36.
22. The technique was first proposed in Pearson, K. (1901) On lines and planes of closest fit to a system of points in space, *Phil. Mag.*, **2**, 557–72; a feasible computational solution was first described in Hotelling, H. (1933) Analysis of a complex of statistical variables into principal components, *J. Ed. Psych.*, **24**, 417–41 and 498–520.
23. Peel, C., Willis, M.J. and Tham, M.T. (1992) Enhancing feedforward neural network training. Chapter 4 in this book.
24. Chen, S., Billings, S.A. and Grant, P.M. (1990) Non-linear system identification using neural networks. *Int. J. Control*, **51**(6), 1191–1214; Chen, S., Cowan, C.F.N., Billings, S.A. and Grant, P.M. (1990) Parallel recursive prediction error algorithm for training layered neural networks. *Ibid*, 1215–28.
25. Qin, S.-Z., Su, H.-T. and McAvoy, T.J. (1992) Comparison of four neural net learning methods for dynamic system identification. *IEEE Trans. Neural Networks*, **3**(1), 122–30.
26. Willis, M.J., Di Massimo, C., Montague, G.A., Tham, M.T. and Morris, A.J. (1991) Artificial neural networks in process engineering. *IEE Proc. Pt. D*, **138**(3), 256–66.
27. Klopf, A.H. (1982) *The Hedonistic Neuron: A Theory of Memory, Learning, and Intelligence*. Hemisphere, Washington, D.C.
28. Miller III, W.T., Sutton, R.S. and Werbos, P.J. (1990) *Neural Networks for Control*. MIT Press, Cambridge, Mass.
29. Evans, J.T., Gomm, J.B., Williams, D., Lisboa, P.J.G. and To, Q.S. (1992) A practical application of neural modelling and predictive control. Chapter 6 in this book.
30. Dubois, D. and Prade, H. (1980) *Fuzzy Sets and Systems: Theory and Applications*. Academic Press, New York.
31. Narendra, K.S. and Parthasarathy, K. (1990) Identification and control of dynamical systems using neural networks. *IEEE Trans. Neural Networks*, **1**(1), 4–27.
32. Wu, Q.H., Hogg, B.W. and Irwin, G.W. (1992) A neural network regulator for turbogenerators. *IEEE Trans. Neural Networks*, **3**(1), 95–100.
33. Widrow, B. (1986) Adaptive inverse control. *Proc. IFAC Conference on Adaptive Systems in Control and Signal Processing*, Lund, Sweden, 1–5.
34. Duda, R.O. and Hart, P.E. (1973) *Pattern Analysis and Scene Classification*. Wiley, New York.

Identification of linear systems using recurrent neural networks 3

D.T. Pham and X. Liu

3.1 ABSTRACT

This chapter describes the application of recurrent neural networks of the type proposed by Elman to the identification of linear dynamic systems. It was found that in their basic form Elman nets were only capable of modelling first-order systems. A modification was made to the net structure enabling systems of higher orders to be modelled. Tests have been carried out on systems of up to the third order and satisfactory results have been obtained.

3.2 INTRODUCTION

There are two main kinds of neural networks: feedforward networks and recurrent networks. Feedforward networks do not have feedback connections while recurrent networks have.

A number of drawbacks are associated with using feedforward networks for system identification [1, 2], including slow computation and difficulty in obtaining an independent system simulator [3, 4].

As will be discussed later in the chapter, recurrent networks do not suffer from the above drawbacks. Among the available recurrent networks, the Elman net [5] has the simplest architecture. It can furthermore work with the standard back-propagation learning algorithm. Although much effort has been spent in applying this net to speech processing, little work has been done in investigating its feasibility in general dynamic systems modelling. This paper presents the results of applying the net to the modelling of linear dynamic systems. It has been shown by simulation that an Elman net is

26 Identification of linear systems

suitable for modelling only first-order linear dynamic systems. Based on this observation, a modification to the net has been proposed. The modified Elman net has been used successfully to model dynamic systems of up to the third order. Studies were also conducted to compare the tapped delay line method and the modified Elman net. It was found that the former was less sensitive to the network architecture and learning parameter values but had slow convergence, and the latter was sensitive to learning parameters but converged faster when they had been correctly selected. The main body of the chapter contains two sections. Section 3.3 presents a brief description of the basic Elman net. Section 3.4 discusses the modified Elman net and the modelling results obtained using this net.

3.3 BASIC ELMAN NET

A block diagram of an Elman net is shown in Fig. 3.1. From this, it can be seen that in an Elman net, in addition to the input units, hidden units and output units, there are also context units. The input and output units interact with the outside environment, while the hidden and context units do not. The input units are only buffer units which pass the signals without changing them. The output units are linear units which sum the signals fed to them. The hidden units have linear/non-linear activation functions. The context units are used only to memorize the previous activations of the

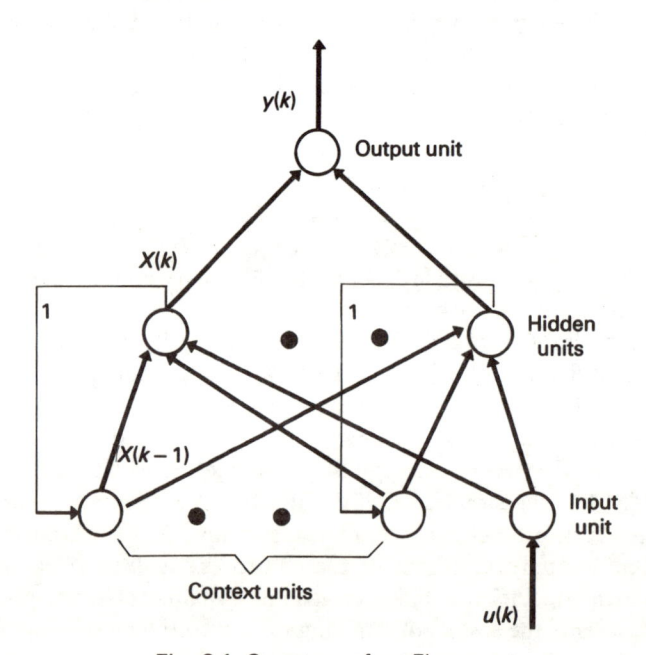

Fig. 3.1 Structure of an Elman net.

hidden units and can be considered to function as one-step time delays. The feedforward connections are modifiable; the recurrent connections are fixed. Because the recurrent connections are fixed, the Elman net is only partially recurrent.

At a specific time k, the previous activations of the hidden units (at time $k-1$) and the current inputs (at time k) are used as inputs to the network. At this stage the network is a feedforward network. These inputs are propagated forward to produce the outputs. The standard back-propagation learning rule [6] is then employed to train the network. After this training step, the activations (at time k) of the hidden units are sent back through the recurrent links to the context units and saved there for the next training step (time $k+1$). At the beginning of training, the activations of the hidden units are unknown. Thus, they are usually set to one half of the maximum range of the value that the hidden units can take. For a sigmoidal activation function, the initial values can be set to 0.5, and for a hyperbolic tangent function, they can be set to 0.0.

3.4 DYNAMIC SYSTEMS MODELLING USING ELMAN NETS

3.4.1 MODIFIED ELMAN NET

Linear Elman nets were used to identify six linear dynamic systems of orders ranging from first to third. During computer simulations, it was found that the Elman nets did not seem to be able to learn the behaviour of a dynamic system higher than the first order. For this reason, an idea based on the work of [7] was employed to configure a modified Elman net which is shown in Fig. 3.2. The modified Elman net differs from the basic Elman net by having self-feedback links with fixed gain α in the context units. Thus the output of the context units can be described by:

$$\text{output}(k) = \text{activation}(k-1) + \alpha \times \text{output}(k-1) \qquad (3.1)$$
$$\text{(context)} \qquad \text{(hidden)} \qquad \text{(context)}$$

When the gain α is zero, the modified Elman net is identical to the original Elman net.

3.4.2 SIMULATION OF DYNAMIC SYSTEMS

A program was written in C on an IBM 386 compatible personal computer to simulate the operation of the modified Elman net. In all the simulations, a string of uniform random inputs u was sent to the systems to produce a training data set of 400 points. The network was trained with the structure of Fig. 3.3. The same structure was employed for recall. The training results were evaluated using the r.m.s. errors between the responses from the system

28 Identification of linear systems

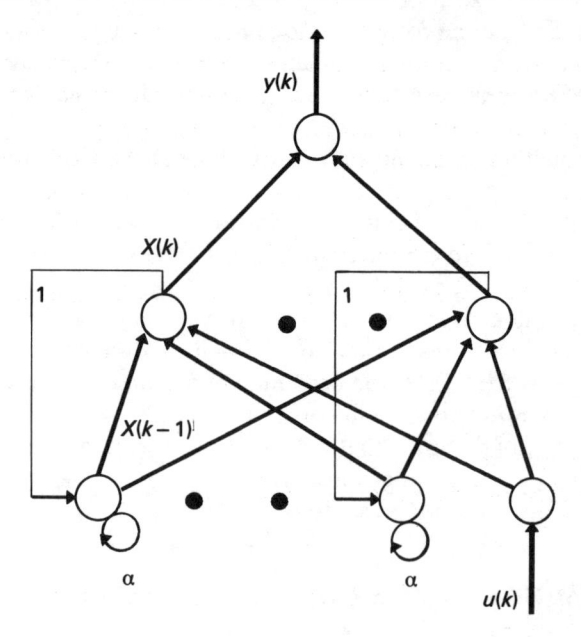

Fig. 3.2 A modified Elman net.

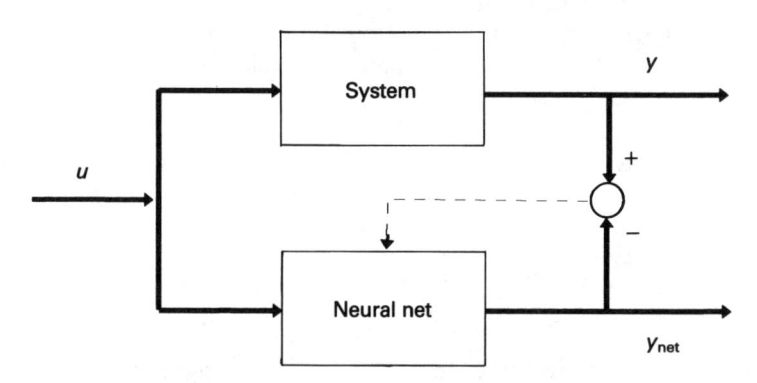

Fig. 3.3 Learning structure.

and the network to a step input ($u(k)$=constant). One hundred data points were used to calculate r.m.s. errors, namely:

$$\text{rms error} = \left(\frac{\sum_{k=1}^{100} [y(k) - y_{\text{net}}(k)]^2}{100} \right)^{1/2} \tag{3.2}$$

Three groups of simulations were conducted. The first group tests the effects of various αs, the second group, the effects of different initial weights and the third group, the influence of different input training signals (note that different initial weights and different input signals were obtained by employing different values for the seeds of the random number generator). Simulations were concentrated on gain α, initial weights and training signals because it was found that these factors influenced the results significantly.

3.4.3 SYSTEM MODELS USED IN SIMULATIONS

The systems to be modelled were all linear sampled-data systems of orders ranging from 1 to 3.

First-order systems

- A pure integrator.
- A general first-order system with pole:

$$p = -1.0$$

Second-order systems

- A second-order system with two real poles:

$$p_1 = -1.0; \quad p_2 = -2.0$$

- A second-order system with two complex conjugate poles:

$$p_{1,2} = -1.0 \pm j2\pi/2.5.$$

Third-order systems

- A third-order system with three real poles:

$$p_1 = -1.0; \quad p_2 = -2.0; \quad p_3 = -3.0.$$

- A system with one real pole and two complex conjugate poles:

$$p_1 = -2.5; \quad p_{2,3} = -1.0 \pm j2\pi/2.5.$$

3.4.4 SIMULATION RESULTS AND DISCUSSIONS

The results for the pure integrator are presented in Table 3.1(a). Table 3.1(a) shows that $\alpha = 0.0$ gives the best result among all the αs. As expected, both initial weights and input signals affect the training, with the input having the most influence. The number of hidden units was varied between 1 and 8. No general trend regarding the speed of learning could be observed in this respect. The results presented in Table 3.1(a) were obtained with two hidden units.

30 Identification of linear systems

Table 3.1(a) Data for the integrator

seed(w) = 1, seed(u) = 2		seed(u) = 2, a=0.0		seed(w) = 1, a=0.0	
a	r.m.s. error	seed(w)	r.m.s. error	seed(u)	r.m.s. error
0.0	0.025164	1	0.025164	1	4.644600
0.1	1.376147	2	0.051386	2	0.025164
0.2	2.408331	3	0.241440	3	0.129860
0.3	3.412959	4	0.023541	4	5.813700
0.4	4.033041	5	0.212329	5	5.810648
0.5	4.407331	6	0.436581	6	5.784981
0.6	4.967193	7	0.059046	7	0.026907
0.7	5.053646	8	0.040771	8	5.751942
0.8	overflow				
0.9	overflow				

hidden units: 2; learning rate: 0.1; momentum: 0.1; sampling period: 0.1; training iterations: 20 000; training input: $u \in [-1.0, +1.0]$; recall input: $u=1.0$.

Table 3.1(b) Data for the first-order system

seed(w) = 1, seed(u) = 2		seed(u) = 2, a=0.0		seed(w) = 1, a=0.0	
a	r.m.s. error	seed(w)	r.m.s. error	seed(u)	r.m.s. error
0.0	0.003283	1	0.003283	1	0.014615
0.1	0.026816	2	0.003107	2	0.003283
0.2	0.047600	3	0.004271	3	0.001959
0.3	0.057867	4	0.003176	4	0.005389
0.4	0.058450	5	0.008669	5	0.001416
0.5	0.049067	6	0.002181	6	0.003042
0.6	0.032151	7	0.000822	7	0.003539
0.7	0.024721	8	0.002843	8	0.007503
0.8	0.023444				
0.9	0.036559				

hidden units: 2; learning rate: 0.1; momentum: 0.1; sampling period: 0.1; training iterations: 20 000; training input: $u \in [-1.0, +1.0]$; recall input: $u=1.0$; $p=-1.0$.

The results for the general first-order system are shown in Table 3.1(b). Table 3.1(b) indicates that again $\alpha=0.0$ produces the best result. Initial weights and input signals also affect the training operation as in the case of the pure integrator.

Table 3.2(a) shows the results for the second-order system with two real poles. During simulations, it was noticed that the training results changed considerably with the value of α (ranging from 0.0 to 0.9). $\alpha=0.5$ produced the best result in terms of the shape of the network response curve, although when $\alpha=0.6$ the recall r.m.s. error was the smallest. Based on this observation, with α fixed to 0.5, different initial weights and input signals were tested. It was found that when seed(w) (the seed for the generator of random initial weights) was 1 and seed(u) (the seed for the generator of random input

Dynamic systems modelling using Elman nets 31

Table 3.2(a) Data for the second-order system (two real poles)

seed(w) = 1, seed(u) = 1		seed(u) = 1, a = 0.5		seed(w) = 1, a = 0.5	
a	r.m.s. error	seed(w)	r.m.s. error	seed(u)	r.m.s. error
0.0	0.455099	1	0.244557	1	0.244557
0.1	0.515530	2	0.537470	2	0.018245
0.2	0.528856	3	0.182639	3	0.013470
0.3	0.503694	4	0.277781	4	0.027397
0.4	0.430671	5	0.159912	5	0.079681
0.5	0.244557	6	0.358708	6	0.028641
0.6	0.164375	7	0.252264	7	0.067639
0.7	0.271790	8	0.658147	8	0.156649
0.8	0.388254				
0.9	0.869532				

hidden units: 3; learning rate: 0.1; momentum: 0.1; sampling period: 0.2; training iterations: 100 000; training input: $u \in [-1.0, +1.0]$; recall input: $u = |p_1 p_2|$; $p_1 = -1.0$; $p_2 = -2.0$.

signals) was 3 the network gave responses which were closest to the system to be modelled. Moreover, it was observed that the learning speed was dependent on the number of hidden units used and sensitive to input signals and initial weights. Several tests were also conducted when gain $\alpha = 0.0$, with pole a fixed and pole b variable. It was found that the r.m.s. error decreased as b was increased. This shows that the original Elman net tends to perform better when the system degenerates to the first order.

For the second-order system with two complex poles, observations similar to those for the second-order system with two real poles were recorded. The results are presented in Table 3.2(b).

Table 3.2(b) Data for the second-order system (two complex poles)

seed(w) = 1, seed(u) = 1		seed(u) = 1, a = 0.6		seed(w) = 1, a = 0.6	
a	r.m.s. error	seed(w)	r.m.s. error	seed(u)	r.m.s. error
0.0	0.380383	1	0.090172	1	0.090172
0.1	0.663787	2	0.873257	2	0.056384
0.2	0.853222	3	0.989033	3	0.075684
0.3	0.906353	4	0.750705	4	0.070921
0.4	0.843345	5	0.111723	5	0.042278
0.5	0.579901	6	0.419687	6	0.078080
0.6	0.090172	7	0.539494	7	0.054033
0.7	0.234413	8	0.186885	8	0.100238
0.8	0.294819				
0.9	0.382749				

hidden units: 4; learning rate: 0.1; momentum: 0.1; sampling period: 0.1; training iterations: 40 000; training input: $u \in [-1.0, +1.0]$; recall input: $u = (1 + \omega^2)/\omega$; $\omega = 2\pi/2.5$; $p_{1,2} = -1 \pm j\omega$.

32 Identification of linear systems

The results for the third-order system with three real poles are shown in Table 3.3(a). It can be seen that $\alpha = 0.65$ yields the best response. With $\alpha = 0.65$, different initial weights and input signals were used and the data obtained show that different initial weights and input signals produce different recall r.m.s. errors when the learning iteration number is fixed. This means that the convergence speed again depends on the initial weights and input signals.

The results for the third-order system with one real pole and two complex poles are shown in Table 3.3(b). In this case $\alpha = 0.85$ gives the best net-

Table 3.3(a) Data for the third-order system (three real poles)

seed(w) = 1, seed(u) = 1		seed(u) = 1, a = 0.65		seed(w) = 1, a = 0.65	
a	r.m.s. error	seed(w)	r.m.s. error	seed(u)	r.m.s. error
0.0	0.888142	1	0.025685	1	0.025685
0.1	0.888815	2	0.216637	2	0.068922
0.2	0.889304	3	0.416218	3	0.062481
0.3	0.889418	4	0.039441	4	0.019314
0.4	0.888816	5	0.031232	5	0.053267
0.5	0.886308	6	0.883726	6	0.094862
0.6	0.114019	7	0.386195	7	0.027279
0.65	0.025685	8	0.178488	8	0.054375
0.7	0.144925				
0.8	0.252444				
0.9	0.140226				

hidden units: 4; learning rate: 0.01; momentum: 0.1; sampling period: 0.1; training iterations: 200 000; training input: $u \in |p_1 p_2 p_3|$; recall input: $u = p_1 p_2 p_3$; $p_1 = -1.0$; $p_2 = -2.0$; $p_3 = -3.0$.

Table 3.3(b) Data for the third-order system (one real and two complex poles)

seed(w) = 1, seed(u) = 1		seed(u) = 1, a = 0.85		seed(w) = 1, a = 0.85	
a	r.m.s. error	seed(w)	r.m.s. error	seed(u)	r.m.s. error
0.0	0.769284	1	0.014848	1	0.014848
0.1	0.773149	2	0.008791	2	0.011529
0.2	0.751731	3	0.097009	3	0.012984
0.3	0.677190	4	0.157557	4	0.016806
0.4	0.535100	5	0.019677	5	0.020695
0.5	0.350345	6	0.030501	6	0.012006
0.6	0.144499	7	0.067910	7	0.018687
0.7	0.045902	8	0.027311	8	0.023381
0.8	0.083530				
0.85	0.014848				
0.9	0.224983				

hidden units: 4; learning rate: 0.05; momentum: 0.1; sampling period: 0.08; training iterations: 200 000; training input: $u \in |p_1|(1+w^2)$; recall input: $u = |p_1|(1+w^2)$; $w = 2\pi/2.5$; $p_{2,3} = -1.0 \pm jw$.

work. Again it is apparent that the recall r.m.s. errors vary with different initial weights and input signals when the learning iteration number is fixed.

As an example, the step response of the third-order system with one real pole and two complex poles and the response of the modified Elman net are depicted in Fig. 3.4.

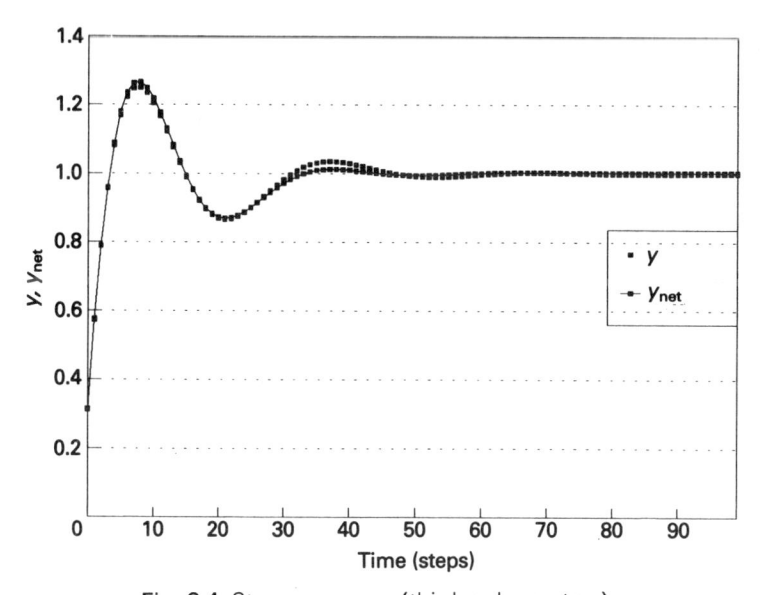

Fig. 3.4 Step responses (third-order system).

From the above description, it can be seen that although the Elman net has been considered capable only of recognizing sequences or producing short continuations of known sequences [1], it is also able to model dynamic systems. The original Elman net can model only first-order dynamic systems. After modification, the net can model dynamic systems of up to the third order in the simulations carried out by the authors. Because an identical modified Elman net structure was used for the second- and third-order systems, it is believed that that net can model a dynamic system of arbitrary order. Due to the systems and networks used in this work, the above conclusion is limited to linear systems.

According to the simulation results, although it was difficult to achieve good training with a hidden unit number exactly equal to the order of the system to be modelled, having one or two additional units greatly expedited the convergence. On the other hand, according to the theory of system identification [8], a good model can only be achieved by estimating the states and parameters at the same time. It is the authors' view that this is why good results could not be achieved for higher order systems with the basic Elman net.

The last point noticed from the simulations is that the value of α is important for good training. For different systems, there are different suitable values of α. For the work reported in this paper α was changed manually. New algorithms or new network achitectures are needed to enable a network to find the proper value of α by itself.

3.5 CONCLUSION

Linear Elman nets and a modified version of these nets have been employed to identify linear systems. These nets have been tested on a variety of systems of up to the third order and successful results have been obtained.

ACKNOWLEDGEMENT

The authors would like to thank the CVCP for supporting Xing Liu's studies.

REFERENCES

1. Hertz, J., Krogh, A. and Palmer, R.G. (1991) Introduction to the theory of neural computation. In *Recurrent networks*. Addison-Wesley, Calif., Chapter 7.
2. Williams, R.J. (1990) Adaptive state representation and estimation using recurrent connectionist networks. In *Neural Networks for Control*, (ed. W.T. Miller III, R.S. Sutton and P.J. Werbos). MIT Press, Cambridge, Mass., pp. 97–114.
3. Pham, D.T. and Liu, X. (1990) State-space identification of dynamic systems using neural networks. *Engineering Applications of Artificial Intelligence*, 3, 198–203.
4. Pham, D.T. and Liu, X. (1991) Neural networks for discrete dynamic system identification. *J. of Systems Engineering*, 1(1), 51–60.
5. Elman, J.L. (1990) Finding structure in time. *Cognitive Science*, 14, 179–211.
6. Rumelhart, D.E. and McClelland, J.L. (1986) *Parallel Distributed Processing*, Vol. 1. MIT Press, Cambridge, Mass., Chapter 8.
7. Jordan, M.I. (1986) Attractor dynamics and parallelism in a connnectionist sequential machine. *Proc. of the Eighth Annual Conf. of the Cognitive Society*, Amherst, Mass., 531–46.
8. Eykhoff, P. (1974) *System Identification: Parameter and State Estimation*. John Wiley, London.

Enhancing feedforward neural network training

4

C. Peel, M.J. Willis and M.T. Tham

4.1 ABSTRACT

A new paradigm for enhancing the training of feedforward artificial neural networks is described. The technique utilizes a polynomial approximation to the sigmoidal processing function [1] and directly integrates principal component analysis (PCA) into the network training philosophy. A major benefit of the new technique is that off-line network training is 'one-shot' as opposed to the usual iterative techniques cited in the literature. Further training may be performed recursively, yielding an adaptive neural network. Additionally, the new philosophy incorporates a systematic procedure for determining the number of neurons in the hidden layer of the network.

This contribution is organized as follows. The training procedure is first described and the implications of the training philosophy discussed. Some results, including applications to industrial chemical processes, are then presented to highlight the power of the technique. The systems considered are a continuous stirred tank reactor and a polymerization reactor.

4.2 INTRODUCTION

Within the chemical/process engineering fraternity, it has been suggested that a reason for the current strong interest in artificial neural networks (ANNs) is due to their ability to represent non-linear systems [2-4]. The ability of multi-layer feedforward neural networks to approximate general mappings from one finite dimensional space to another has been demonstrated by many applications to various systems in the field of science, engineering and information processing [5]. Thus, ANNs have the potential of providing a generic 'cost effective' tool for developing models of notoriously complex chemical/biochemical processes. It is generally acknowledged that the development of realistic mechanistic models of such systems is costly, both in terms of time and effort.

In the literature many application results assess and demonstrate the benefits of an ANN approach. However, many issues remain unresolved [3]. The problem addressed in this contribution is that of speeding up the process of network training. The proposed procedure tackles this via the use of polynomial-based non-linear processing functions and the incorporation of a mechanism for the automatic selection of the most appropriate input data to use for model building.

Due to the popularity of the non-linear processing functions, such as the sigmoidal function, radial basis functions and so on, the algorithms used to train artificial neural networks are based on iterative searches. Thus, training can be time-consuming. However, it has been shown that the use of polynomial-based non-linear functions in neural networks is just as effective [1]. By specifying polynomials which are linear in the parameters, non-iterative procedures such as the well-known least-squares technique may be employed. Iterative network training can therefore be reduced to a 'one-shot' procedure. Moreover, if the recursive forms of the least-squares parameter estimation algorithm are used, then network training can be carried out on line.

Training speed is also influenced by the number of nodes in the network and it is suggested that the number of nodes necessary is determined by the complexity of the relationships between input and output data. A technique which can be used to assess the relative contribution of the inputs to the output variations is that of principal component analysis (PCA). PCA may be classified as a multivariate data analysis technique and is an invaluable procedure for the ratiocination and interpretation of complex multivariable data sets. It provides a powerful means for highlighting the main direction of data variations in a high-dimensional space. Hence, in performing a PCA, the analysis of a large number of variables may be reduced to the investigation of a smaller transformed set. It has been demonstrated that the philosophy facilitates data validation and fault detection [6] as well as process quality control [7, 8]. PCA is related to partial least squares (PLS) regression, and a comparison of the utility of the neural network approach and PLS technique was presented in McAvoy et al. [9]. Their results showed that a feedforward artificial neural network (FANN) was superior to PLS in terms of its estimation ability. Nevertheless, the desire to incorporate information from a PLS analysis into the neural network modelling procedure resulted in the formulation of a generic non-linear PLS approach using neural networks [10]. This technique decomposes the multi-input multi-output non-linear modelling task into linear 'outer', and non-linear 'inner' relations performed by a number of single-input single-output networks. Following decomposition of the network, a conjugate gradient algorithm is employed to train the

FANNs. Although similar concepts are adopted, the technique proposed in this contribution is significantly simpler and allows substantially more insight into network behaviour. Moreover, via the direct incorporation of PCA into the FANN training philosophy, it is possible to specify the (minimum) number of neurons required in the hidden layer of the network in a more formal manner. The smaller the number of network nodes, the fewer parameters there are to determine. Hence, network training speed will be enhanced.

4.3 FEEDFORWARD ARTIFICIAL NEURAL NETWORKS

Whilst a number of ANN architectures have been proposed [5], the 'feedforward' ANN (FANN) is by far the most widely applied. A possible explanation for this could be due to the results of Cybenko [11] and Hornik et al. [12]. They have shown that a FANN with two layers of non-linearity is sufficient to approximate, arbitrarily well, any continuous non-linear function. Hence, this will be the topology adopted.

A FANN is composed of layers of interconnected non-linear processing elements, often called nodes, as denoted by the circles in Fig. 4.1. Each connection has an associated weight which acts to modify the signal strength. Scaled data is fed into the network at the input layer. The signals are then 'fed forward' through the network along the connections to the output layer. The 'neurons', except those in the input layer, perform two functions. They act as a summing point for the input signals as well as propagating the input through the non-linear processing function. Generally, FANNs utilize the following non-linear 'sigmoidal' processing function:

$$\sigma(x_j) = \frac{1}{1 + e^{-x_j}} \quad (4.1)$$

where $\sigma(x_j)$ = the output of unit j.

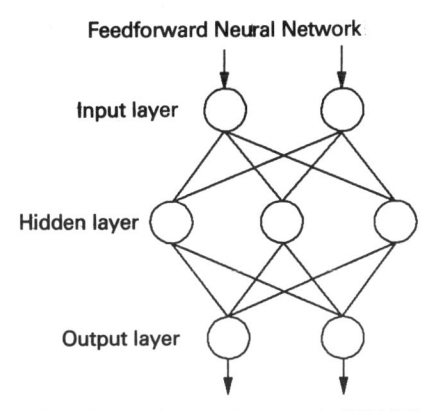

Fig. 4.1 Typical feedforward neural network (FANN) architecture.

In developing a neural network model, the topology (structure) of the network must first be declared. Once the network topology is specified, the network is 'trained', i.e. a set of system input and output data is used to determine the appropriate values of the network weights (including the bias terms) associated with each interconnection. Essentially, the training procedure searches for an extremum (maximum or minimum) of a real function f of n variables (the network weights) where the objective function is typically specified as:

$$f(w) = \sum_{i=1}^{p} E_i \qquad (4.2)$$

where $E_i = (y_d - y)^2$, y is the output of the network and y_d is the target or desired output of the network.

Although the minimization of the error, E_i, can be achieved by a number of minimization procedures, the back-error propagation algorithm [13] is commonly applied. Back-error propagation is based upon one of the more primitive optimization techniques, known as the method of steepest descent. Recently there have been several attempts to improve the back-propagation algorithm. It should be noted that the steepest descent method, and hence back-propagation, locally linearizes the function $f(\mathbf{w})$ by fitting a tangent plane to the objective function. Thus, except for that contained in the slope, all information about the function is lost when this minimization technique is used. The problem may be alleviated by the application of second-order optimization procedures, i.e. those that utilize the second partial derivatives of $f(\mathbf{w})$ with respect to the network weights.

An obvious drawback with second-order methodologies, especially for large networks, is the need to compute second-order derivatives. Thus, whilst convergence properties may be improved, considerably more calculations per epoch would be required. The usual recourse is therefore to approximate the Hessian matrix, leading to the use of quasi-Newton techniques. Watrous [14] evaluated Davidon–Fletcher–Powell (DFP) and Broyden–Fletcher–Goldfarb–Shanno (BFGS) variants and compared these to back-error propagation with an optimized learn rate. It was shown that both the DFP and BFGS algorithms took fewer iterations than back-error propagation, but at the expense of higher computational overheads due to the Hessian updates. As a result, Watrous [14] concluded that second-order methods may not be advantageous for large networks.

In another development, Leonard and Kramer [15] suggested the use of a conjugate directions search algorithm to improve convergence rate, whilst minimizing the computational burden. Although quasi-Newton methods usually converge more rapidly and are generally more robust than conjugate gradient methods, they require significantly more storage. Another advantage of the conjugate gradient method is that second-order convergence properties can be achieved without having to calculate the second derivatives of the objective function.

Although the above-mentioned alternatives to back-propagation have led to faster network training, they are nonetheless iterative procedures. As a result they are still subject to the problem of local minima and are slower than 'one-shot' techniques. In this contribution, a new non-iterative approach for network training is described. The number of neurons in the hidden layer is also automatically determined as part of the procedure, a decided improvement over trial and error approaches to network topology selection (as often appears to be the case in other published evaluation studies). The topology selection procedure is based on the concepts used in principal component analysis.

4.4 PRINCIPAL COMPONENT ANALYSIS

Consider an example where there are m observation of n variables. This data set is presented as an $(m \times n)$ matrix. Each column of the matrix is then standardized to yield the matrix $\mathbf{X}$. Thus, each column of $\mathbf{X}$, $(\mathbf{x}_i, i=1,\ldots,n)$ represents a sequence of observations of the ith variable transformed to have zero means and unit variances. The matrix $\mathbf{X}$ can then be decomposed as:

$$X = U\Sigma^{1/2}V^T \tag{4.3}$$

where $\mathbf{U}$ is an $(m \times m)$ matrix of the eigenvectors of $\mathbf{XX}^T$ and $\mathbf{V}$ is an $(n \times n)$ matrix of the eigenvectors of $\mathbf{X}^T\mathbf{X}$, the data correlation matrix. The elements of the diagonal matrix $\Sigma^{1/2}$ are the positive square roots of the eigenvalues, $\lambda_i(i=1,\ldots n)$ of $\mathbf{X}^T\mathbf{X}$ and are called 'singular values'. The columns of $\mathbf{U}$ and $\mathbf{V}$ are termed the left and right 'singular vectors' of $\mathbf{X}$ respectively, and they have the following properties:

$$\mathbf{V}^T\mathbf{V} = \mathbf{V}\mathbf{V}^T = \mathbf{I} \quad \text{and} \quad \mathbf{U}^T\mathbf{U} = \mathbf{U}\mathbf{U}^T = \mathbf{I}.$$

Note that the matrix decomposition technique described by equation (4.3), called the singular value decomposition (SVD), is merely one of the numerous methods for determining the eigenvectors and eigenvalues of matrices and is therefore not fundamental to PCA. However, it is perhaps the most numerically robust procedure for accomplishing this task [16] and is also applicable to non-square matrices. In practice, this is useful as the amount of data available is usually more than the number of variables under consideration.

The principal components of $\mathbf{X}$ are the columns of the matrix $\mathbf{P}$ defined as:

$$\mathbf{P} = XV = U\Sigma^{1/2}. \tag{4.4}$$

Thus, for an $(m \times n)$ data matrix, there will be n principal components. The ith principal component, $\mathbf{p}_i$, is essentially a weighted sum of the standardized variables, where the weights are defined by the elements of the ith eigenvector, $\mathbf{v}_i$, i.e.:

$$p_i = Xv_i \quad (i=1,\ldots,n). \tag{4.5}$$

Note that equation (4.5) can also be expressed as a linear combination of the columns of X in the following manner:

$$p_i = \sum_{j=1}^{n} v_{ij} x_i \quad (i=1,\ldots,n) \tag{4.6}$$

where $v_{i,j}$ is the jth element of the ith eigenvector of the correlation matrix $X^T X$. Therefore, from equation (4.6), it can be deduced that $v_{i,j}$ indicates the extent to which the ith principal component depends on the jth variable. From equation (4.5), the variance of $\mathbf{p}_i$ is given by:

$$\text{var}(p_i) = v_i^T X^T X v_i = \lambda_i, \tag{4.7}$$

i.e. the variance of each principal component is given by the corresponding eigenvalue, λ_i, of the data correlation matrix [17–19]. Since the correlation matrix is symmetric, all its eigenvalues have magnitudes greater than or equal to zero. They can also be arranged in descending order. In this case, the first principal component, corresponding to the largest eigenvalue, will explain most of the variation in the given data. The second principal component explains the cause of the next level of variation, and so on. By discarding those principal components which do not contribute significantly to overall data variations, the dimension of the problem is correspondingly reduced. It is the purpose of this chapter to demonstrate that the technique can be beneficially used as an integral part of the neural network training procedure.

4.5 APPLICATION OF PRINCIPAL COMPONENT ANALYSIS TO NEURAL NETWORKS

As discussed earlier, it has been shown that a FANN with only two layers of non-linearities can be used to approximate any continuous function arbitrarily well. The weights in these two layers are usually determined using a minimization routine in order to obtain the necessary degree of approximation. However, in this novel methodology to be described, the determination of the weights in the two layers differs from that in the normal FANN architecture. The weights on the connections between the input layer and the hidden layer are fixed by the results of a PCA on the input data. Thus, as discussed above, the variables taken as inputs to the network are gathered in the X matrix and a principal component analysis is performed on the data correlation matrix $X^T X$, i.e. the original data is transformed in the manner shown in equation (4.3).

The resulting principal components are linearly uncorrelated with each other even though the original variables may exhibit multi-collinearities. Since the matrix of principal components is obtained by multiplying the input matrix $\mathbf{X}$ by $\mathbf{V}$ (the matrix of eigenvectors of $\mathbf{X}^T\mathbf{X}$), the weights between the input and hidden layer are taken as the elements of the corresponding eigenvectors. The input data to the net are therefore transformed by the first set of weights, i.e. the signals to the hidden layer are principal components and hence are not correlated. Furthermore, the percentage contribution of each principal component to the overall variation in the data may then be determined. Those components that do not offer any significant explanation may be eliminated from further analysis. Within the present context, this translates to the elimination of unnecessary nodes in the hidden layer. Thus, an additional benefit of the proposed procedure is that the number of neurons in the hidden layer is also specified.

A possible drawback of the approach is that the weights of the bias terms associated with the hidden layer are not specified using the PCA. It has been shown [1] that networks without bias are unable to approximate certain functions. To elucidate, consider the Taylor series expansion of a nodal transfer function without bias, equation (4.1):

$$f_a(x) = \frac{1}{1+e^{-x_j}} = a_0 + a_1 x_j + a_2 x_j^3 + a_3 x_j^5 + \cdots. \tag{4.8}$$

In other words, a network without bias cannot generate a series with even powers of inputs. Hence the network may be unable to achieve the necessary degree of approximation of certain function types. However by the addition of bias into the nodal transfer function the following series expansion may be written:

$$f_b(x) = \frac{1}{1+e^{-(x_j+b)}} = a_0 + a_1 x_j + a_2 x_j^2 + a_3 x_j^3 + \cdots, \tag{4.9}$$

i.e. a polynomial with both odd and even powers. Thus, the bias term yields a network which has greater flexibility and capability to model non-linear functions.

Unfortunately, the weights in the bias layer cannot be determined from the PCA analysis and must be determined separately. An alternative implementation would be to use PCA as a network initialization procedure. On having established a set of weights using PCA (rather than using a randomized set of initial weights), the network may then be trained using any standard technique, such as back-propagation. It is suggested that this would enhance the speed of network training. However, in this contribution we wish to develop a one-shot training technique. In order to achieve this objective, it is necessary to assume that the network processing functions in the hidden layer can be approximated by equation (4.9), while the output

42 Enhancing feedforward neural network training

non-linearity remains as the sigmoidal processing function. This approximation allows the following equation to be written:

$$y'(t) = \sum_{h=1}^{N} w_h \sum_{j=0}^{k} s_{j,h} u_h(t)^j \qquad (4.10)$$

where: $s_{j,h} = j$th coefficient of the hth polynomial neuron in the hidden layer; w_h = weight from hth neuron to the output neuron; $u_h(t)$ = total input to the hth neuron in the hidden layer at time t; and $y'(t)$ = the desired output data passed through the inverse of the output node non-linearity at time t. Note that in order to perform this inversion it is assumed that the output data is scaled in the range [0.1, 0.9]. N is the number of neurons in the hidden layer; k is the neuron polynomial order.

Equation (4.10) may be simplified by multiplication of the respective w_h into the coefficients $s_{j,h}$ of the respective polynomials to yield:

$$y'(t) = \sum_{h=1}^{N} \sum_{j=0}^{k} s'_{j,h} u_h(t)^j \qquad (4.11)$$

where $s'_{j,h}$ represents the transformed polynomial coefficients which must be determined such that the desired accuracy of approximation is achieved. During network training a batch of input-output data is used to determine the coefficients of the network. If it is assumed that the training data set comprises data from time $t = 1, \ldots, p$ then, equation (4.11) may be rewritten in the following matrix form:

$$y = X\Theta \qquad (4.12)$$

where:

$$y = [y'_j(t), y'_j(t+1), \ldots, y'_j(t+p)]^T$$

$$\Theta = [s_{0,1}, s_{1,1}, \ldots, s_{k,1}, \ldots, s_{0,N}, s_{1,N}, \ldots, s_{k,N}]^T$$

$$X = \begin{bmatrix} 1 & u_1(t) & \cdots & u_1(t)^k & \cdots & 1 & u_N(t) & \cdots & u_N(t)^k \\ \vdots & \vdots & \ddots & \vdots & & \vdots & \vdots & \ddots & \vdots \\ 1 & u_1(t+p) & \cdots & u_1(t+p)^k & \cdots & 1 & u_n(t+p) & \cdots & u_n(t+p)^k \end{bmatrix}$$

The solution of equation (4.12) allows the determination of the coefficients of the polynomials. Further, the numerically robust SVD procedure can also be used to perform this linear regression, i.e. an iterative technique to train the network is not necessary.

4.5.1 OPTIMUM TOPOLOGY SELECTION

It has been shown that the non-iterative procedure developed above may be used to train the FANN. However, in order to obtain the network topology that will yield the optimum representation of the non-linear system, an iterative cross validation technique should be used. The proposed technique can be summarized by the following steps.

Step 1: Split the data into at least two representative sets.
Step 2: Take a single set of data as the training data set. Perform the PCA on this data. This allows:
  (a) the weights between the input layer and the hidden layer to be determined;
  (b) the elimination of C_f neurons in the hidden layer. This is due to the fact that the PCA allows the calculation of the percentage contribution of each principal component to the overall variation in data. With this information it is possible to eliminate the components that do not offer any signficant explanation of the data variations. In the present context, this is equivalent to the elimination of unnecessary nodes in the hidden layer.
Step 3: Choose a low polynomial order, k_p, for the neurons in the hidden layer. Note that, for simplicity during application studies, it is assumed that the order of the polynomial in each neuron is the same. However, this is not a necessary assumption.
Step 4: Train the network using linear least squares. Evaluate the prediction error on the test data set.
Step 5: Increment k_p and go to Step 3, until $k_p > k_{pmax}$, where k_{pmax} is the maximum polynomial order to be considered.
Step 6: Decrease C_f and go to Step 2(b).
Step 7: Obtain the value of k and N (where N = number of neurons in the hidden layer) for which the prediction error on the test data set is a minimum.
Step 8: Further test the 'optimal' model on an unseen data set.

This simple cross-validation procedure therefore allows the 'optimal' polynomial order as well as the 'optimal' number of neurons to be specified for the network. The network will be 'optimal' in terms of the minimum integral square error (ISE) between actual process output and the predicted network output on the test data set. Since the minimization of the objective function on the training data set is achieved in a 'one-shot' manner, the above procedure is computationally efficient.

4.6 DISCUSSION

The network training technique discussed above has been developed for cases where multi-collinearities in the data are assumed to exist. However, it is suggested that the technique is also applicable to the training of all network models. If multi-collinearities are not present (or in a limiting case where there is only one network input) then the weight(s) in the input layer could be fixed to one: it is the coefficients of the respective polynomials $(s'_{j,h})$ that must be determined such that the desired accuracy of approximation is achieved.

Another point to note is that the approximation problem is reduced to that of a linear regression problem. Thus recursive least squares [20] can be utilized to train the neural network. This allows the development of self-tuning/adaptive neural networks. The obvious advantage of this is that if process conditions change, the model parameters can be adjusted on line to accommodate for such time-varying behaviour.

4.7 SELF-TUNING/ADAPTIVE NETWORKS

The self-tuning/adaptive neural network training procedure utilizing recursive least squares (RLS) may be summarized as follows:

Step I: Based on a predefined batch of training and test data perform the batch identification procedure (Steps 1–7) as discussed above. This enables the 'optimal' number of neurons and polynomial order to be determined.

Step II: Based on an unseen data set, set up the data vector x, i.e.

$$\mathbf{x} = [1, u_1(t), \ldots, u_1(t)^k, \ldots, 1, u_N(t), \ldots, u_N(t)^k]^T.$$

Use RLS to estimate the parameter vector, $\Theta(t)$:

$$\Theta(t) = [s_{0,1}, s_{1,1}, \ldots, s_{k,1}, \ldots, s_{0,N}, s_{1,N}, \ldots, s_{k,N}]^T$$

such that

$$\frac{\partial(y-\hat{y})^2}{\partial \Theta(t)} = \frac{\partial(y-\mathbf{x}^T\hat{\Theta})^2}{\partial \Theta(t)} = 0.$$

This is process identification step and completes the self-tuning part of the algorithm.

Note that in all the results to be presented, the PCA is not performed on line and hence the weights in the input layer to the network remain fixed. Obviously, this does not have to be the case since the PCA could be continually performed on a moving 'window' of data.

4.8 PERFORMANCE EVALUATION

One of the reasons for increasing industrial interest in the application of artificial neural networks is that they have the potential for improving the quality of on-line information available for plant control and optimization. In the chemical, biochemical and food processing industries major problems exist with regard to the on-line estimation of parameters and variables that quantify process behaviour. Briefly, the problem is that the key 'quality' variables cannot be measured at a rate which enables their effective regulation. This can be due to limited analyser cycle times or a reliance upon off-line laboratory assays. An obvious solution to such problems could be realized by the use of a model along with secondary process measurements, to infer product quality variables at the rate at which the secondary variables are available. Hence, if the relationship between quality measurements and on-line process variables can be captured, then the resulting model can be utilized within a control scheme to enhance process regulation. The following examples demonstrate the capabilities of the new neural network training philosophy.

4.8.1 ESTIMATION OF PRODUCT CONVERSION IN A CONTINUOUS STIRRED TANK REACTOR (CSTR)

The process under consideration is the highly non-linear CSTR studied by Economou and Morari [21]. The objective of control in this system is to operate the reactor at near maximum feed conversion whilst, at the same time, preventing thermal runaway. The manipulative variable is feed temperature while the controlled variable is product concentration. The ultimate purpose of the work is to develop a dynamic network controller [3]. However, before this may be accomplished a suitable neural network model must be developed. Thus, in this preliminary study, a neural network is trained to predict product concentration based on a time history of feed temperatures.

Input–output data from the process is obtained by perturbing the feed temperature by a pseudo-random binary sequence and data was collected at 30 second intervals. The network was trained using five past values of feed temperature measurements as input variables whilst the network output was the current measurement of feed concentration. Following the procedure set out in Steps 1–7 outlined previously, the optimum number of neurons was found to be five, and the processing functions third-order polynomials. The latter was determined based on the ISE obtained during cross-validation of the model using the test data set.

Figure 4.2 shows the results of the cross-validation procedure. For a 5-5-1 network, (i.e. a network with five input neurons, five neurons in the hidden layer and a single output neuron) it may be observed that as the order of the polynomial functions increases, the ISE associated with the training data set decreases. At first sight, the results are indicative of an improvement in the quality of approximation. However, when the model is cross-validated on the test data set, it may be observed that a minimum ISE exists, corresponding to a polynomial order of 3. A possible explanation is that during training, the network is forced to match all the response characteristics of the output signal. Higher order polynomials provide the capability of allowing matching peaks in the response. On the other hand, after training,

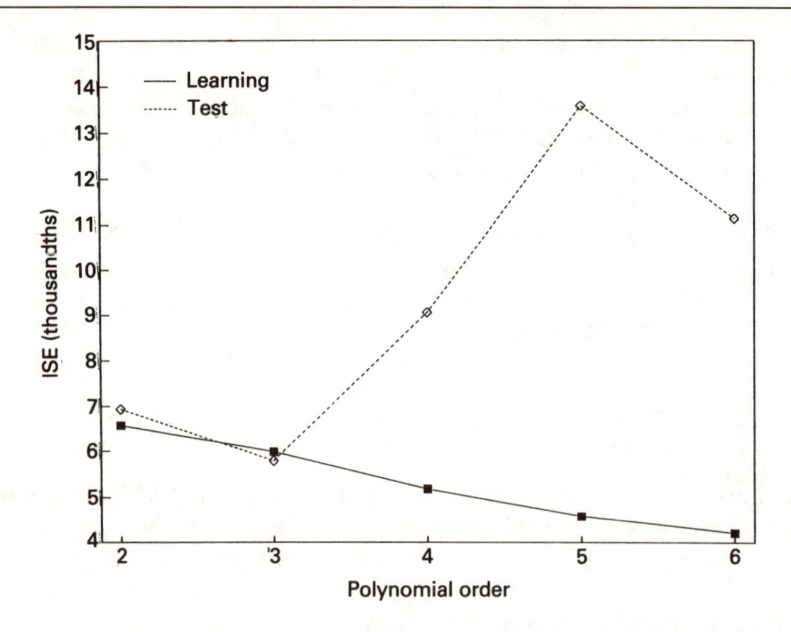

Fig. 4.2 Determination of the optimal polynomial size for a 5-5-1 network.

the network is required to extrapolate at least one step ahead, where the use of order polynomials is very susceptible to errors. Thus, it is the ISE obtained from the test data set which is used as the performance index for determining the necessary number of neurons and the order of the polynomial functions.

Having determined the network topology as 5-5-1 with third-order polynomial functions, i.e. a 5-5(3)-1 polynomial network, the model was then used 'on line' to predict product concentration. The quality of the approximation achieved by the neural network is shown in Fig. 4.3 where the performance of an adaptive version of the corresponding neural network is also presented. The latter made use of the optimum network topology that had been determined 'off line'. The RLS identification was then performed 'on line', to provide the self-tuning capabilities. The initial covariance matrix for the RLS was specified at 1000*I, where I is the identity matrix, and the initial parameter estimates were set to 0.0. A fixed forgetting factor ($\lambda = 0.98$) was used within the RLS algorithm to exponentially discount old data and hence enhance the tracking of non-linear process characteristics. The respective coefficients of the polynomial functions in each neuron are shown in Figs 4.4 to 4.8. It may be observed that after an initial tuning phase, the network weights remain essentially constant, confirming that the non-linear neural network is of sufficient accuracy to approximate the non-linear process. The variations in network weights may be attributed to the changes in system characteristics as the system is perturbed about its operating point.

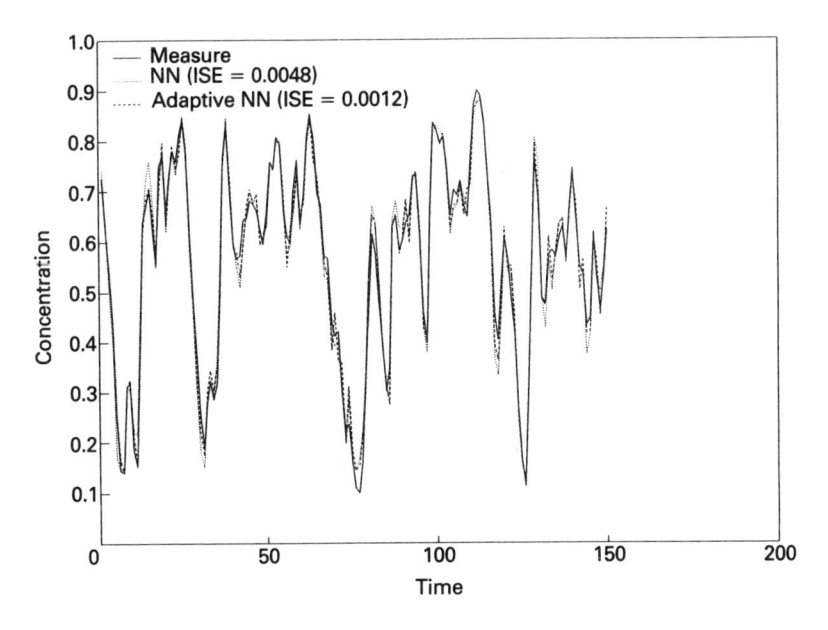

Fig. 4.3 Estimation of product composition in a CSTR.

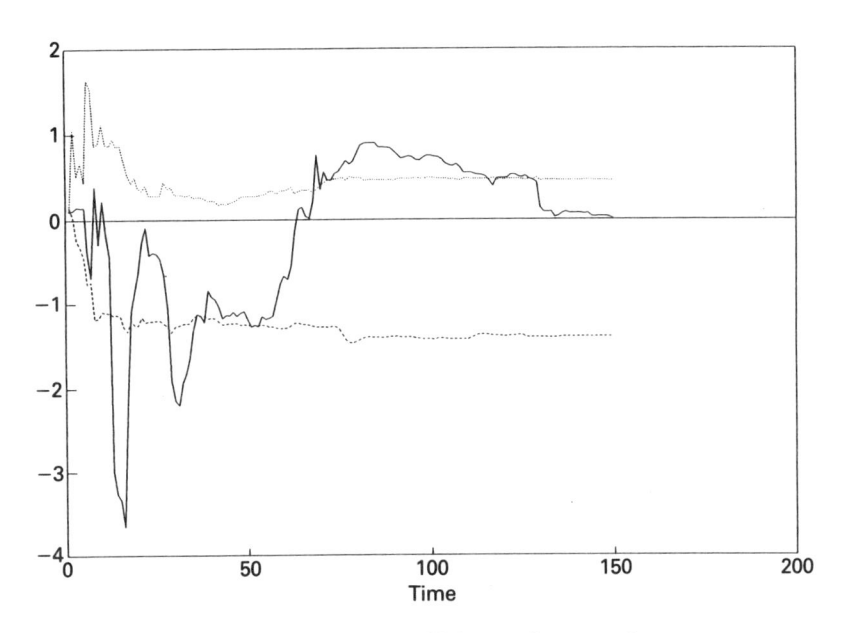

Fig. 4.4 Estimated coefficients of neuron 1.

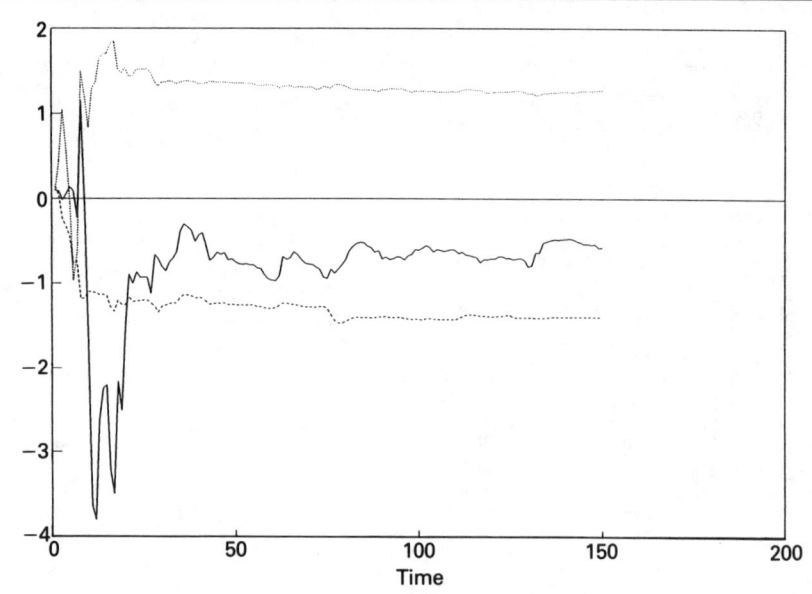

Fig. 4.5 Estimated coefficients of neuron 2.

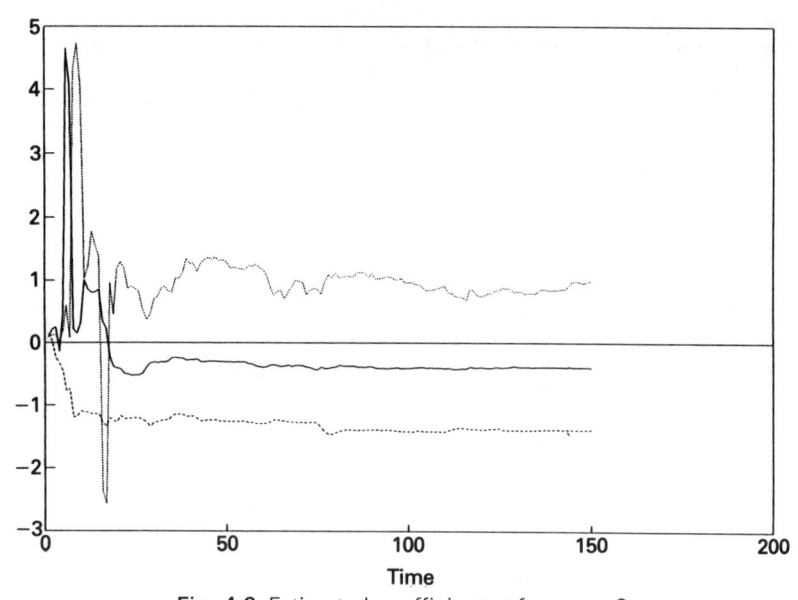

Fig. 4.6 Estimated coefficients of neuron 3.

4.8.2 ESTIMATION OF MELT FLOW INDEX (MFI) IN INDUSTRIAL POLYMERIZATION

In this application, a neural network is required to provide estimates of a measure of product quality, MFI, in a polymerization reactor. Unfortun-

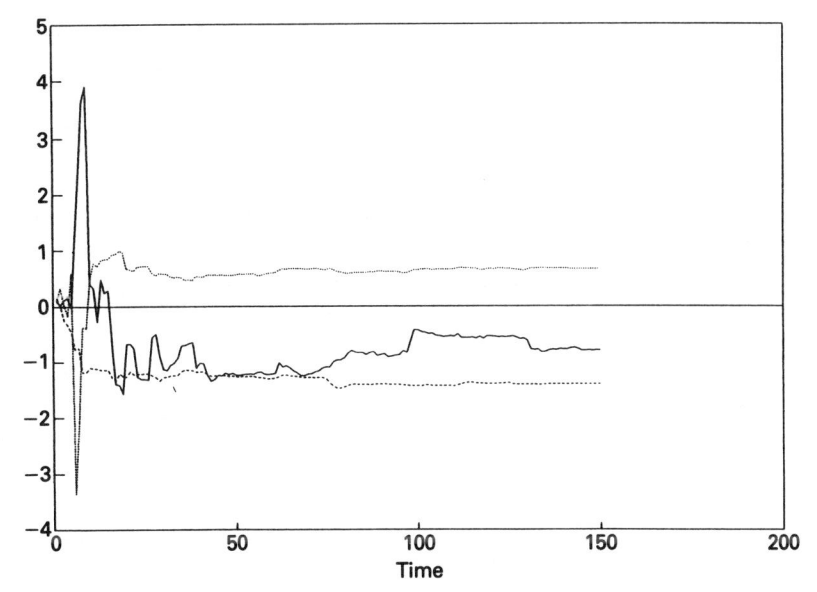

Fig. 4.7 Estimated coefficients of neuron 4.

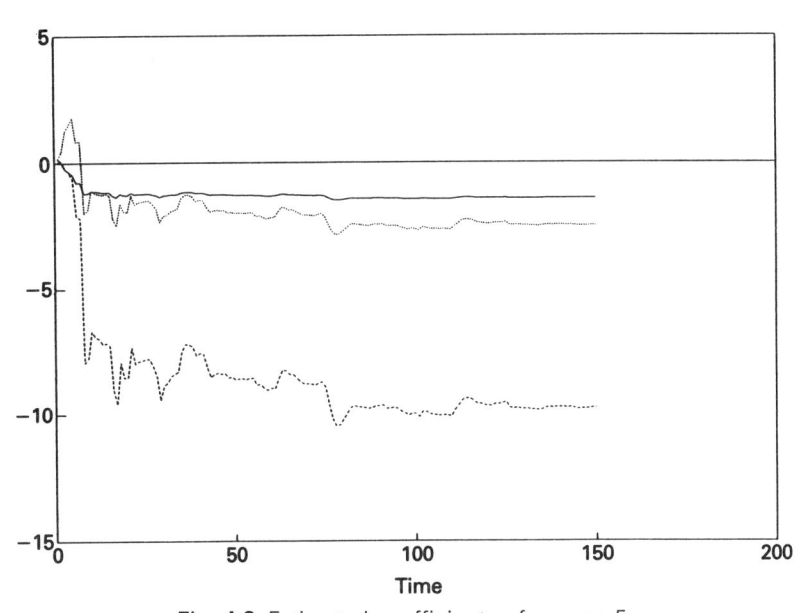

Fig. 4.8 Estimated coefficients of neuron 5.

ately, industral confidentiality precludes a thorough description of the process. Through process analysis, nine other process variables were identified as those that could affect MFI. The measurements of these variables were therefore taken as inputs to the network model while the output was the current measurement of MFI. The batch procedure was used to train the network model and it was determined that the most appropriate network topology was 9-4(4)-1. The quality of prediction using the network is demonstrated in Fig. 4.9. Again, the results from using the adaptive network are presented on the same plot. It may be observed that the non-adaptive network provided relatively poor estimates of MFI. A possible reason for this is that process conditions were changing due to the advent of process disturbances. This therefore presents an ideal example for the use of an adaptive network and the algorithm was thus implemented 'on line' using the *a priori* determined best topology of 9-4(4)-1. As with the previous implementation, the initial convariance matrix for the RLS was specified at 1000*I; a fixed forgetting factor of 0.98 was employed and the parameters initialized to zeros. As illustrated in Fig. 4.9, the adaptive neural network provided significantly improved MFI estimates. The ISE obtained for the non-adaptive network was 0.0084 as compared to 0.0045 for the adaptive version.

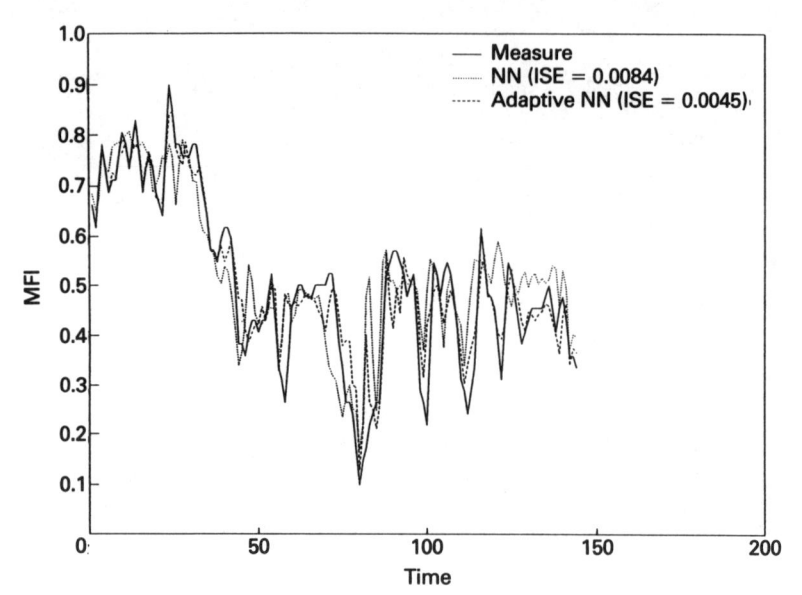

Fig. 4.9 Estimation of MFI in an industrial polymerization process.

4.9 CONCLUDING REMARKS

In this contribution, a new training paradigm for FANNs has been introduced. The technique utilizes a polynomial approximation to the sigmoidal processing function and provides an efficient means of choosing the optimal topology of a single hidden layered FANN. Application to industrial process data revealed that, given an appropriate topology, the network could be trained to characterize the behaviour of the systems considered. Additionally, it was demonstrated that in systems where process characteristics change, an adaptive FANN model may be appropriate as it can improve the degree of approximation.

It should be noted that PCA is a linear analysis technique applied to non-linear data. Moreover, this technique only discriminates between linear correlations in the input data. Such a limitation could be perceived as a potential drawback of the approach. It is conjectured, therefore, that the development of a recursive PCA technique would benefit the technique. This is the subject of current work.

ACKNOWLEDGEMENTS

The authors would like to thank the members of the artificial neural network research groups at the Universities of Newcastle upon Tyne and Maryland and the Industrial Neural Network Club members.

REFERENCES

1. Wray, J. and Green, G.G.C. (1991) Analysis of networks that have learnt control problems. *IEE International Conference CONTROL '91*, Edinburgh, UK.
2. Bhat, N. and McAvoy, T.J. (1990) *Comput. Chem. Eng.*, **15**(5), 573–83.
3. Willis, M.J., Di Massimo, C., Montague, G.A., Tham, M.T. and Morris, A.J. (1991) *IEE Proc. Pt. D*, **138**(3), 256–66.
4. Zupan, J. and Gasteiger, J. (1991) *Analytica Chimica Acta*, **248**, 1–30.
5. Lippmann, R.P. (1987) *IEEE ASSP Magazine*, **April**, 4–42.
6. Wise, B.M. and Ricker, N.L. (1989) Upset and sensor failure detection in multivariate processes. AIChE Meeting, San Francisco, USA.
7. MacGregor, J. (1989) Multivariate statistical methods for monitoring large data sets from chemical processes. AIChE Meeting, San Francisco, USA.
8. Efthimiadu, I. and Tham, M.T. (1991) On-line statistical process control of chemical processes. Preprints of *IEE International Conference CONTROL '91* Edinburgh, UK, 700–5.
9. McAvoy, T.J., Su, H.T., Wang, N.S., He, M. and Horvath, J. (1991) A comparison of neural networks and partial least squares for deconvoluting fluorescence spectra. *Proc. of Second Annual Meeting of the International Neural Networks Club*, Newcastle upon Tyne, UK.
10. Qin, S.J. and McAvoy, T.J. (1991) Neural net PLS approach to dynamic modelling: method and application. Presented at the AIChE Meeting, Chicago, USA.
11. Cybenko, G. (1989) *Math. Cont. Signal and Systems*, **2**, 303–14.
12. Hornik, K., Stinchcombe, M. and White, H. (1989) *Neural Networks*, **2**, 359–66.
13. Rumehart, D.E. and McClelland, J.L. (eds.) (1987) *Parallel Distributed Processing: Explorations in the Microstructure of Cognition*. MIT Press, Cambridge, Mass.
14. Watrous, R.L. (1987) Learning algorithms for connectionist networks: applied gradient methods for nonlinear optimisation. *Proc. Int. Conf. on Neural Networks*, *II*, pp. 619–27.

15. Leonard, J. and Kramer, M.A. (1990) *Comput. Chem. Eng.*, **14**, 337.
16. Golub, G.H. and Van Loan, C.F. (1983) *Matrix Computations*. North Oxford Pubishing Co., Oxford.
17. Manly, B.F.J. (1986) *Multivariate Statistical Methods: A Primer*. Chapman & Hall, London.
18. Piggott, J.R. and Sharman, K. (1986) *Statistical Procedures in Food Research*. Elsevier Applied Science, London, Chapter 6.
19. Rawlings, J.O., (1988) *Applied Regression Analysis: A Research Tool*. Wadsworth and Brooks, California.
20. Ljung, L. and Soderstrom, T. (1983) *Theory and Practice of Recursive Identification*. MIT Press, Cambridge, Mass.
21. Economou, C.G. and Morari, M. (1982) *Ind. Eng. Chem. Process Des. Dev.*, **25**, 403-11.

Estimation of state variables of a fermentation process via Kalman filter and neural network

5

D. Tsaptsinos, N.A. Jalel and J.R. Leigh

5.1 ABSTRACT

During a fermentation process, variables such as product concentration are determined by off-line laboratory analysis, making this set of variables of limited use for control purposes. In this chapter two approaches for modelling the product concentration in the Cyathus Striatus fermentation process are presented. The first is based on a conventional identification approach while the second is based on a neural network technique. It was decided to adopt sequential modelling to represent the process by dividing it into two phases representing growth and the steady state respectively. The comparison of the two approaches showed that the neural networks outperformed the identification approach during learning and produced a comparable performance during operation.

5.2 INTRODUCTION

The aim of industrial biotechnology is to obtain useful metabolic products from biological material. This process involves two main stages; fermentation and product recovery. Fermentation procedures must be developed for the cultivation of micro-organisms under optimal conditions and for the production of desired metabolites or enzymes by the micro-organisms, while product recovery involves the extraction and purification of biological products.

In the fermentation process there are three types of variables: the controlling inputs to the fermenter such as feeds, pH and temperature; the measured outputs in the gas components leaving the fermenters such as oxygen and carbon dioxide; and the state variables, such as biomass, glucose and product concentration.

The problem with the industrial fermentation process is to model the behaviour of unmeasurable state variables and to design a suitable controller for the process. In controlling a fermentation process, the most difficult problem is to determine its current state. This problem can be overcome if a suitable process model can be found, for then an estimator can be used to determine the non-measurable variables and the model parameters [1].

Two approaches for modelling the product of the Cyathus Striatus fermentation process were used. In the first approach, two linked autoregressive (AR) models were derived to describe the process. The first expresses the relation between the inputs and the unmeasurable state variables, while the second is used to express the relation between the state variables and the measured output. The Kalman filter driven from these two AR models was applied to provide an on-line estimation of the state variables. As an extension to this work, an attempt was made to generate a general non-linear model to describe the process. The non-linear equations are derived by integrating the linear AR models. The values for the non-linear matrix coefficients are found by numerical differentiation.

In the second approach, a neural network technique was adopted for on-line estimation of the product concentration, with the neural network taking on the task of both modelling and state estimation. A two-layered, fully connected, feedforward network was used in which learning is achieved by standard back-propagation. The input layer includes four processing elements to which those variables available for on-line measurement are taken as inputs for the network. The output is the product concentration required to be estimated. Since fermentation is a time-varying process, a sequential modelling approach was adopted. In this approach, the fermentation process is divided into two phases: the growth and the steady state phases. Each phase was treated separately in that a model for each was derived. Of course, an important point is to find the change-over point between the two phases. The change-over point could be found by inspection of the data through graphical representation. The model of the state variables of the process were taken to be the combination of the two models. By comparing with the modelling result of the whole process, it was found that the sequential modelling approach achieved better modelling and representation of the process [2]. This is presumably due to the fact that this approach can partially accommodate the non-linearity of the process.

Modelling the fermentation process 55

The Cyathus Striatus fermentation process that has been modelled in this chapter was run at the University of Kaiserslautern. Access to the data was afforded to us via British Council link/DAAD funded research between the Institute of Biosystems (FAL) Braunschweig, the Technical University of Hamburg–Harburg and the University of Kaiserslautern. We would like to acknowledge the co-operation, assistance and the personal involvement of Professor A. Munack, Mr. V. Hass and Mr. R. Boker [3].

5.3 MODELLING THE FERMENTATION PROCESS USING AN IDENTIFICATION TECHNIQUE

5.3.1 THE AR MODEL FOR THE FERMENTATION PROCESS

In this work the identification approach, auto-regressive (AR), is adopted to develop a model for the batch process from the available data. In this approach *a priori* knowledge is not taken into account but the modeller has access to a large supply of data from which model structure, as well as parameter estimates, may be deduced [4].

In the AR model the dynamic system can be conceptually described as in Fig. 5.1, where the system is driven by the input variable $u(t)$ and the disturbance $e(t)$. For this process two models have been derived to describe the process. In the first a model has been derived to express the relation between the inputs, the dissolved oxygen (PO_2) and the oxygen (O_2), and the unmeasurable state variable, product concentration, Fig. 5.2.

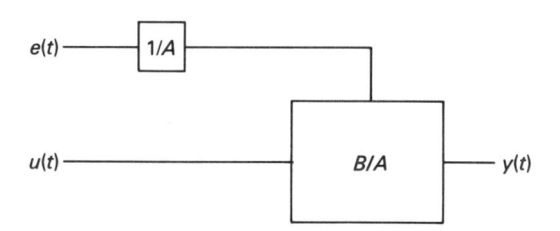

Fig. 5.1 The ARX model.

Fig. 5.2 The two models.

For the first phase, the model after applying AR identification is described in the state space form as

$$\begin{bmatrix} x_1(k+1) \\ x_2(x+1) \end{bmatrix} = \begin{bmatrix} 1.7165 & -0.7136 \\ 1.0 & 0.0 \end{bmatrix} \begin{bmatrix} x_1(k) \\ x_2(k) \end{bmatrix}$$

$$+ \begin{bmatrix} 0.1462 & -0.0329 \\ 0 & 0 \end{bmatrix} \begin{bmatrix} u_1(k) \\ u_2(k) \end{bmatrix}$$

while for the second phase

$$\begin{bmatrix} x_1(k+1) \\ x_2(x+1) \end{bmatrix} = \begin{bmatrix} 1.557 & -0.5701 \\ 1.0 & 0.0 \end{bmatrix} \begin{bmatrix} x_1(k) \\ x_2(k) \end{bmatrix}$$

$$+ \begin{bmatrix} 0.5745 & -0.1161 \\ 0 & 0 \end{bmatrix} \begin{bmatrix} u_1(k) \\ u_2(k) \end{bmatrix}$$

where $x_1(k)$ is the product concentration, $u_1(k)$ and $u_2(k)$ are the O_2 and PO_2 and $x_2(k)$ is a dummy variable introduced to increase the model order. The second model is used to express the relation between the state variable (product) and the measured output CO_2 ($y(k)$), as shown in Fig. 5.2.

The state space equations for the first and second phases have been defined by AR estimation respectively as

$$y(k) = [-0.0019 \quad 0] \begin{bmatrix} x_1(k) \\ x_2(k) \end{bmatrix}$$

$$y(k) = [0.5192 \times 10^{-3} \quad 0] \begin{bmatrix} x_1(k) \\ x_2(k) \end{bmatrix}$$

Here, the change-over point between phases was chosen by inspection of the data at the value $K = 140$.

5.3.2 KALMAN FILTER ESTIMATOR

The most widely used and accepted types of state estimator for both continuous and discrete states are based on the Kalman filter technique. The Kalman filter is a powerful state estimator that accepts measurements corrupted by sensor noise (e.g. noise in the CO_2 measurement) and model noise appearing as errors in the A, B, and C matrices of the AR model and provides an estimate of the inaccessible state variables.

The mathematics of the Kalman filter have been derived in several texts. However, the schematic representation for the Kalman filter is illustrated in Fig. 5.3 [5].

5.3.3 SIMULATION RESULTS

The model has been derived using real data from the fermentation process. Figs 5.4 and 5.5 illustrate the off-line measurements for the product concentration, the model response and the prediction of the Kalman filter estimator for phases I and II respectively.

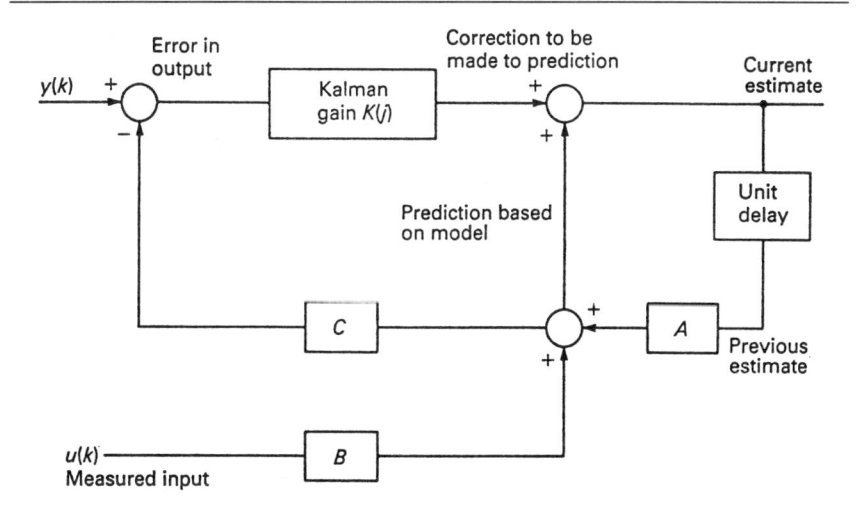

Fig. 5.3 Kalman filter.

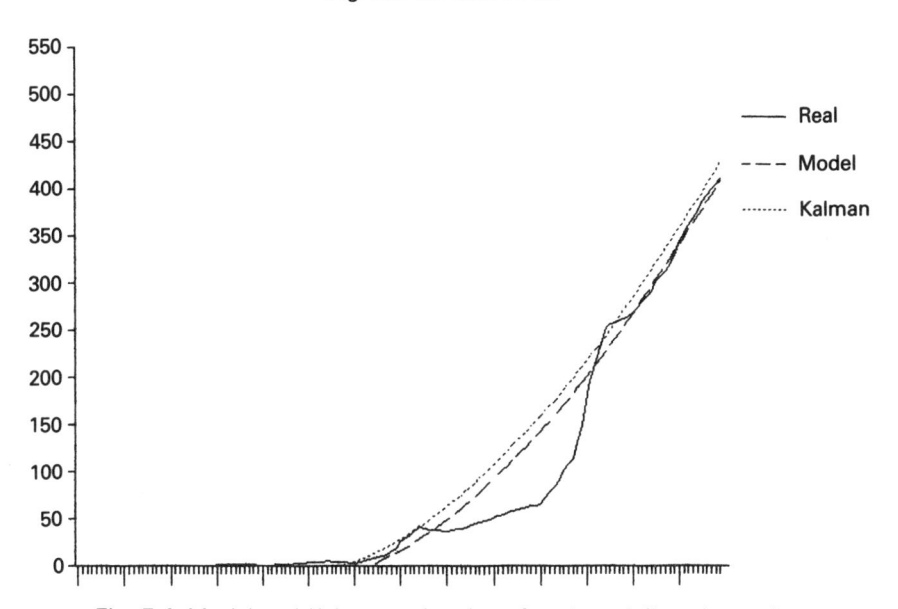

Fig. 5.4 Model and Kalman estimations for phase I (learning set).

To test the ability of the model to estimate the unmeasurable state variables another set of data has been used and applied on the derived model. Figs 5.6 and 5.7 show the real and estimated values of the product versus time for the phases I and II respectively.

From these figures it is worth indicating that a simple AR model could give a good estimation of the product. The comparison of Kalman filter and

58 State variables of a fermentation process

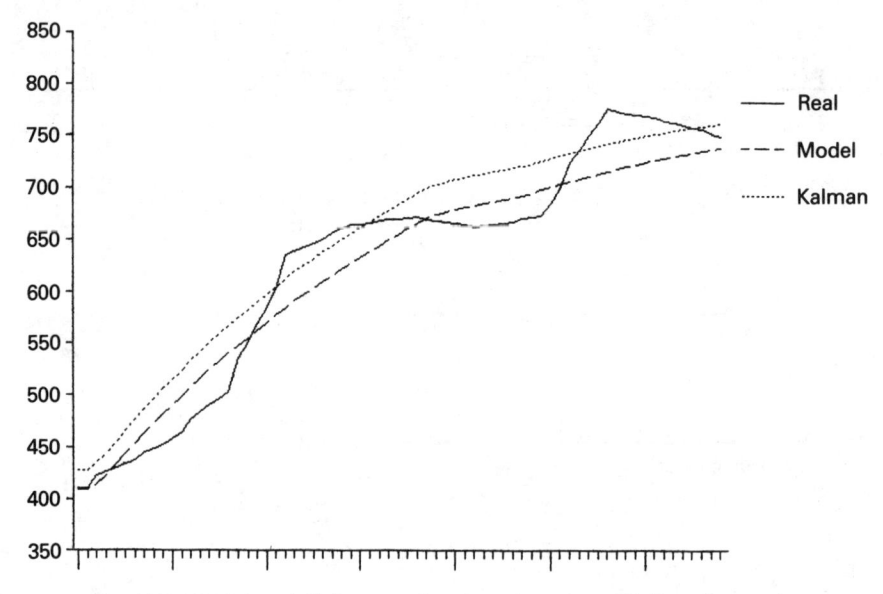

Fig. 5.5 Model and Kalman estimations for phase II (learning set)

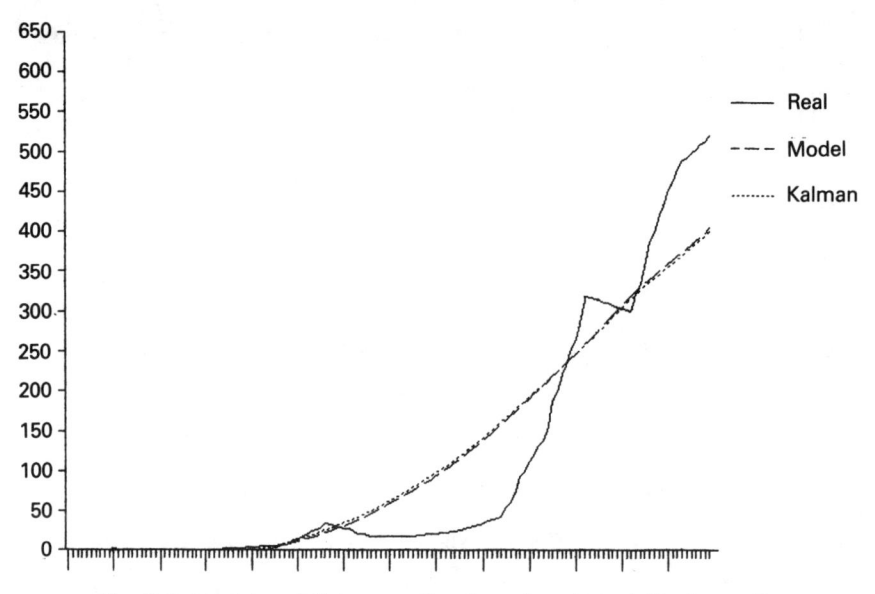

Fig. 5.6 Model and Kalman estimations for phase I (testing set)

model predictions showed a small deviation. This is due to the accuracy of the model prediction in the first place. However in the case of the unseen data, the model prediction in the second phase is not very satisfactory: presumably due to unmodelled batch-to-batch deviations.

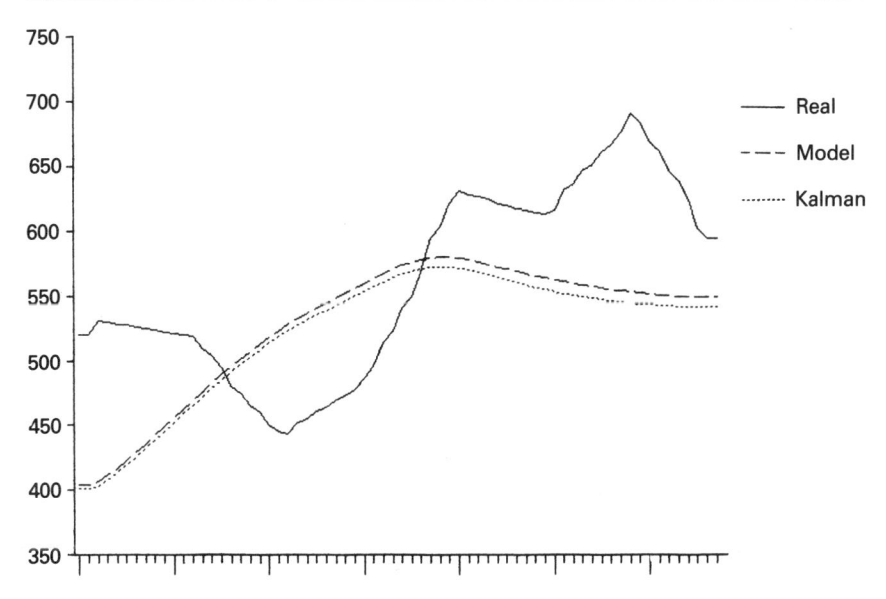

Fig. 5.7 Model and Kalman estimations for phase II (testing set).

5.3.4 NON-LINEAR MODEL OF THE PROCESS

The fermentation process is highly non-linear. In this part of the contribution, an attempt has been made to create a non-linear model of the process. The approach is based on transforming the linear model derived using the AR technique into a non-linear model.

As an example of the procedure for a second-order system, the discrete model is transferred into a continuous form, described as

$$\begin{bmatrix} \dot{x}_1 \\ \dot{x}_2 \end{bmatrix} = \begin{bmatrix} a_{11} & a_{22} \\ a_{21} & a_{22} \end{bmatrix} \begin{bmatrix} x_1 \\ x_2 \end{bmatrix}$$

$$+ \begin{bmatrix} b_{11} & b_{12} \\ b_{21} & b_{22} \end{bmatrix} \begin{bmatrix} u_1 \\ u_2 \end{bmatrix}$$

If this linear model had been derived from a non-linear system containing functions f_1, f_2 then the A and B constants used correspond to

$$\frac{\partial f_1}{\partial x_1} = a_{11}, \quad \frac{\partial f_1}{\partial x_2} = a_{12}, \quad \frac{\partial f_2}{\partial x_1} = a_{21}, \quad \frac{\partial f_2}{\partial x_2} = a_{22} \quad (5.1)$$

$$\frac{\partial f_1}{\partial u_1} = b_{11}, \quad \frac{\partial f_1}{\partial u_2} = b_{12}, \quad \frac{\partial f_2}{\partial u_1} = b_{21}, \quad \frac{\partial f_2}{\partial u_2} = b_{22}. \quad (5.2)$$

60 State variables of a fermentation process

In a truly linear process these values are constant but when they have been derived by linearization from a non-linear model, they should be updated for each sample.

From (5.1) and (5.2)

$$f_1(x_1, x_2) \cong a_{11}x_1 + a_{12}x_2 + b_{11}u_1 + b_{12}u_2 \tag{5.3}$$

$$f_2(x_1, x_2) \cong a_{21}x_1 + a_{22}x_2 + b_{21}u_1 + b_{22}u_2. \tag{5.4}$$

In order to find the updating constant used in the non-linear model, equations (5.3) and (5.4) are differentiated by using the numerical approach which is based on the five-point formula given below

$$f'(x_0) = \frac{1}{12h}[f(x_0 - 2h) - 8f(x_0 - h) + 8f(x_0 + h) - f(x_0 + 2h)] + \frac{h^4}{30}f^5(\zeta). \tag{5.5}$$

By numerical differentiation, values were found for the coefficients a_{ij}, b_{ij}.

5.3.5 NON-LINEAR RESULTS

The above algorithm has been tested using the AR model for the two phases. Figure 5.8 shows the combined model output and the off-line measurement of the product. Again the algorithm has been tested on another set of data and Fig. 5.9 illustrates the combined estimated value of the product concentration. From these figures it can be seen that a good on-line

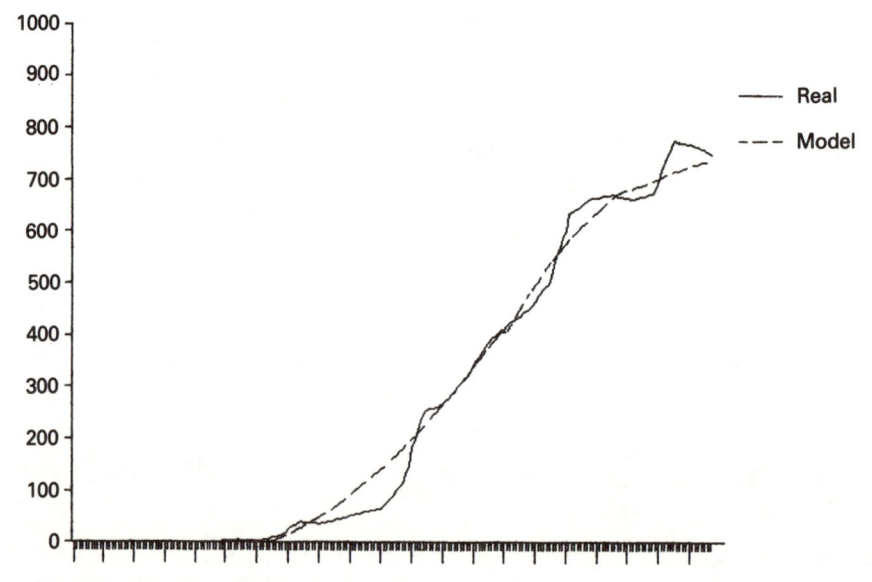

Fig. 5.8 New model estimations for whole process (learning set).

Modelling and state estimation 61

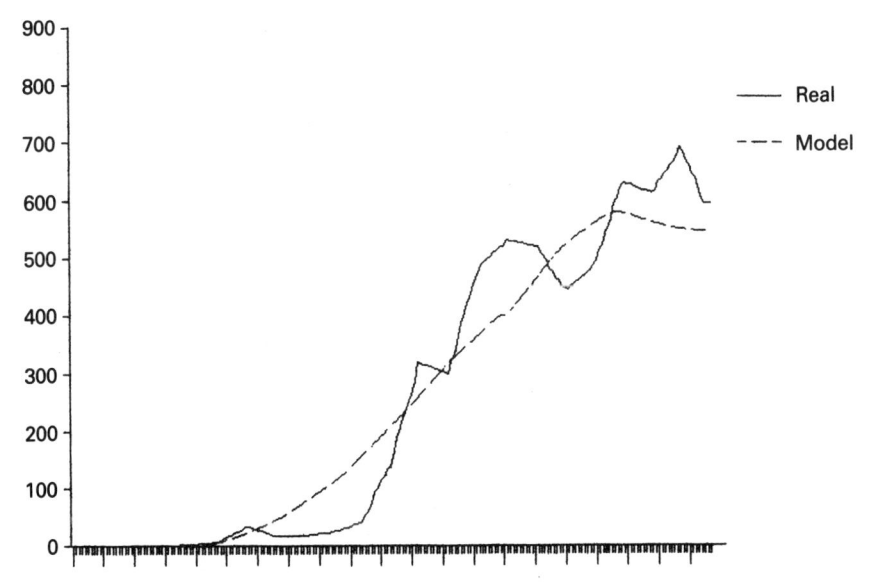

Fig. 5.9 New model estimations for whole process (testing set).

prediction and a non-linear modelling of the product is achieved. Comparing these results with the linear AR model there is not a large deviation. However, the non-linear approach is general and could cope with and model different cases of the process and therefore be more universally applicable.

5.4 MODELLING AND STATE ESTIMATION USING ARTIFICIAL NEURAL NETWORKS

5.4.1 THE FEEDFORWARD NEURAL NETWORK

Artificial neural networks have been inspired by the way the human brain operates but they do not represent exact replicas. A number of neural networks have been reported in the literature [6] and although models differ in detail, each one contains the same basic unit, i.e. the processing element. Processing elements within a neural network are grouped together to form a structure called a layer and a typical hetero-associative neural network usually consists of one (or more) intermediate (alternatively called hidden) layers together with an input and an output layer. Each processing element is fully connected to all elements of the next layer (Fig. 5.10). The function of the input layer is to receive data from the external world and then to pass it to the intermediate layer which in turn processes the data and sends it to the output layer. The output layer then produces a pattern of activation which is compared with the desired pattern of activations to identify if there are any differences. If differences exist the search for the set of weights that will

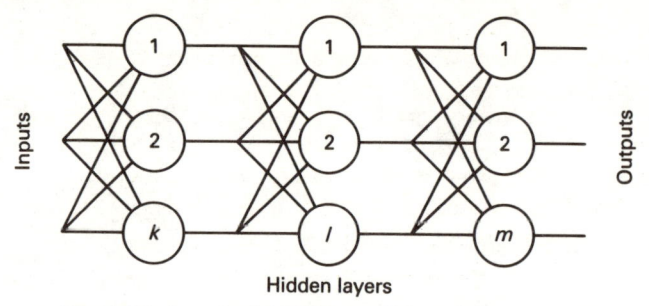

Fig. 5.10 A typical feedforward neural architecture.

produce the desired behaviour commences. The search used in this contribution is known as the back-propagation learning algorithm [7].

5.4.2 THE PREPARATION PHASE: I

A number of questions have to be addressed prior to a neural network implementation. During the preparation phase, one is concerned with what inputs and outputs to employ, the way to present them to the network, the number of nodes per layer, the connectivity pattern, etc. Four inputs were fed to the input layer representing time, oxygen (O_2), dissolved oxygen (DO_2), and carbon dioxide (CO_2). The output of the network is taken to be the product concentration which needs to be estimated on line. The inputs and outputs were scaled between zero and one using the linear transformation of Value−Min. Value/Max. Value−Min. Value. The scaling improved the network performance especially of the second phase.

5.4.3 THE PREPARATION PHASE: II

Selection of the number of processing elements for the input and output layers does not present any difficulties since the numbers are determined by the user and naturally depend on the application. When considering the hidden layers, the selection is more problematic. The selection of how many processing elements to use is typically determined through experimentation. Here, we start with a larger network than we expect to need and using correlation analysis those redundant processing elements and/or connections are taken out. The trimming hopefully leads to a thinner network with the same or better performance than the original one.

5.4.4 STOP-LEARNING CRITERION

During the learning phase, questions arise regarding how long to train the network, when to stop the training and which criteria to use. In the

Modelling and state estimation 63

literature a number of suggestions have been made for when to suspend the learning process, either permanently or temporarily, in order to avoid overspecialization. Here, the following strategy was followed.

1. The maximum acceptable error per outcome for every example of the training set was set. The error refers to the difference of the desired output from the output generated by the network. In this work and for outputs with values scaled between zero and one the individual error was set equal to 0.06. Therefore for a desired outcome with a value of 0.5 a generated output with a value between 0.44 and 0.56 was considered correct. In such cases the weights of the network were not adjusted.
2. The maximum total sum of squared errors (MTSSE) was calculated. 'Total' refers to the error over all examples of the training set. Consequently MTSSE is the product of the number of examples in the training set and the squared individual error. For the first phase MTSSE is equal to 0.504 (140 examples times $(0.06)^2$) and for the second phase is equal to 0.2484 (69 examples times $(0.06)^2$).
3. At intervals, the learning process was interrupted and the performance progress criterion was measured. This criterion involved the calculation of the total sum of squared errors (TSSE). At the first interval where TSSE was less than MTSSE the learning was suspended and the post-pruning technique described in the next section commenced.

5.4.5 USING CORRELATION ANALYSIS TO REDUCE THE NETWORK TOPOLOGY

The correlation analysis is applied in order to establish the level of dependency of the connections between layers. The significance of the inputs to a layer is given by the form of an R_{xy} ($n \times m$) matrix, where n is the number of incoming connections to a layer including the bias connection and m is the number of units of the receiving layer. R_{xy} is obtained using

$$R_{xy} = R_{xx} \times W$$

where W is a weight matrix with dimensions ($n \times m$) and R_{xx} is the auto-correlation matrix and is obtained using

$$R_{xx} = X^T \times X$$

where X is a ($k \times n$) matrix where k is the number of examples of the training set and n, m are as before. X^T denotes the transpose of the X matrix. It is useful to normalize R_{xx} to obtain diagonal elements of value one and also to

64 State variables of a fermentation process

normalize R_{xy} by dividing every element in the matrix by the magnitude of the largest element.

5.4.6 POST-PRUNING PROCESSES I AND II

A network with four inputs, one hidden layer with two processing elements and one output was used originally for both phases. The learning sets for the first and second phases contained 140 and 69 examples respectively. The TSSE was smaller than MTSSE after 90 000 presentations for the first phase and after only 10 000 presentations for the second phase. Tables 5.1, 5.2, 5.3 and 5.4

Table 5.1 Weights on input to hidden layer connections (phase I)

	Hidden node 1	Hidden node 2
Bias	−0.8165	−0.0056
Input node 1	−15.9142	−36.0808
Input node 2	−0.0661	0.3307
Input node 3	−0.6280	2.4471
Input node 4	2.5182	1.7163

Table 5.2 Weights on hidden to output layer connections (phase I)

	Output node 8
Bias	4.9688
Hidden node 6	−6.0016
Hidden node 7	−12.1239

Table 5.3 Weights on input to hidden layer connections (phase II)

	Hidden node 6	Hidden node 7
Bias	−1.1171	−0.6496
Input node 2	1.3329	5.1434
Input node 3	0.6790	−2.5142
Input node 4	−0.1751	0.3757
Input node 5	−1.6574	−1.0926

Table 5.4 Weights on hidden to output layer connections (phase II)

	Output node 8
Bias	−3.5709
Hidden node 6	1.18851
Hidden node 7	5.7843

Modelling and state estimation 65

show the generated weights between the input-hidden and hidden-output layers for phases I and II respectively. Tables 5.5, 5.6, 5.7 and 5.8 show the symbolic correlation matrices as determined with reference to the first four tables. Correlations with values less than 0.5 are represented using a dash, correlation values higher than 0.9 are represented with an asterisk and the question mark represents the remaining values. Connections represented with a dash are then eliminated. Figure 5.11 displays the reduced network topology for the first phase. Figure 5.12 displays the reduced network topology for the second phase.

Table 5.5 Symbolic correlation matrix for the input to output layer connections (phase I)

	Hidden node 6	Hidden node 7
Bias	–	?
Input node 2	–	*
Input node 3	–	?
Input node 4	–	?
Input node 5	–	*

Table 5.6 Symbolic correlation matrix for the hidden to output layer connections (phase I)

	Output node 8
Bias	?
Hidden node 6	?
Hidden node 7	*

Table 5.7 Symbolic correlation matrix for the input to hidden layer connections (phase II)

	Hidden node 6	Hidden node 7
Bias	–	–
Input node 2	–	*
Input node 3	–	?
Input node 4	–	–
Input node 5	?	–

Table 5.8 Symbolic correlation matrix for the hidden to output layer connections (phase II)

	Output node 8
Bias	?
Hidden node 6	*
Hidden node 7	*

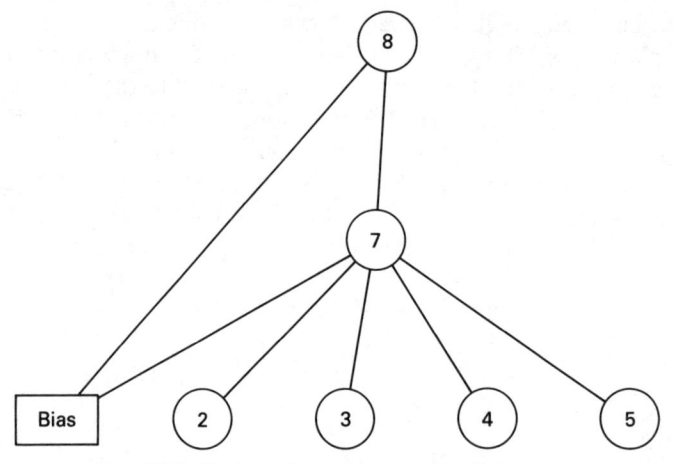

Fig. 5.11 Reduced network topology for phase I.

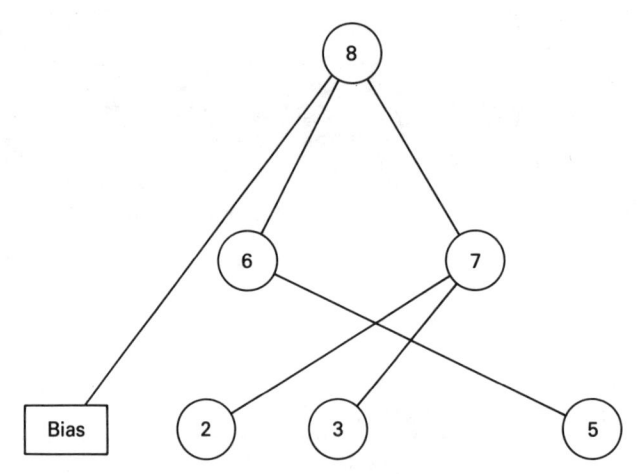

Fig. 5.12 Reduced network topology for phase II.

5.4.7 IMPROVING THE PERFORMANCE

One would like to improve the performance even more. To accomplish this the first FCN network with TSSE smaller than MTSSE was pruned using correlation analysis, and the evolved weights were kept. By assuming that the network has learned the best it can, the initial individual error was halved. The MTSSE for phase I was now 0.126 and for phase II 0.0621. The networks started the learning process and for the first phase it took 50 000 additional presentations to obtain a TSSE value (0.066) less than MTSSE. Unfortunately, it was not possible to achieve TSSE values less than MTSSE for the second phase. This might be due to limitations of the present data.

Modelling and state estimation 67

5.4.8 RESULTS WITH THE LEARNING DATA

Figures 5.13 and 5.14 show the desired and neural network predictions versus time for phases I and II respectively. From both figures it can be concluded that the neural network does a good job in learning both phases. Both sets of predictions were taken from the reduced networks shown in Figs 5.11 and 5.12.

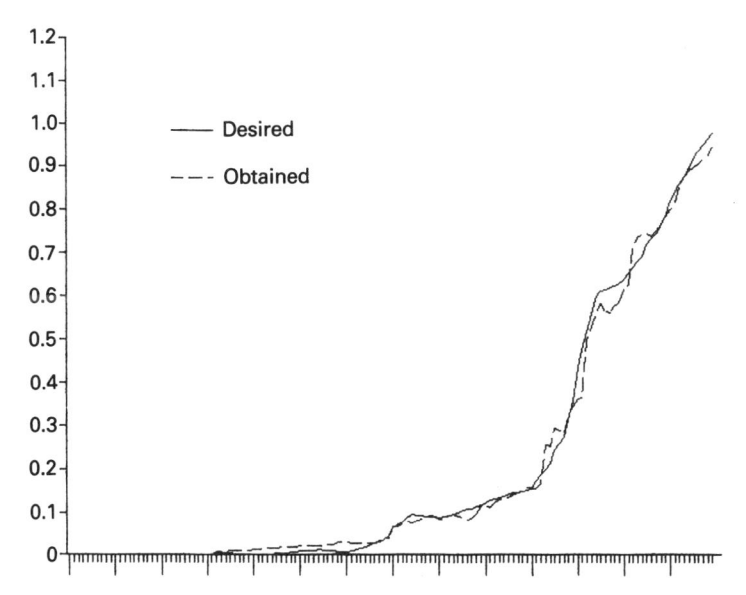

Fig. 5.13 Neural network estimations for phase I (learning set).

5.4.9 RESULTS WITH UNSEEN DATA

The reduced networks were then tested with fermentation data not seen before. Figure 5.15 illustrates the network predictions for the first phase. The predictions of the neural network when applied for the second phase are given in Fig. 5.16. Whereas for phase I the predictions are satisfactory, the same statement cannot be concluded for the second phase. Figure 5.17 shows the product concentration curve of the learning and the testing set for the second phase. As can be seen, the curves are dramatically different, therefore some (or even all) the blame has to be assigned to the data set rather than to the network. This also demonstrates that neural networks have problems with extrapolation, hence in future work extra care must be taken to ensure the validity of any data set and also to ensure that the learning set accommodates not only the normal but also the rare situations.

68 State variables of a fermentation process

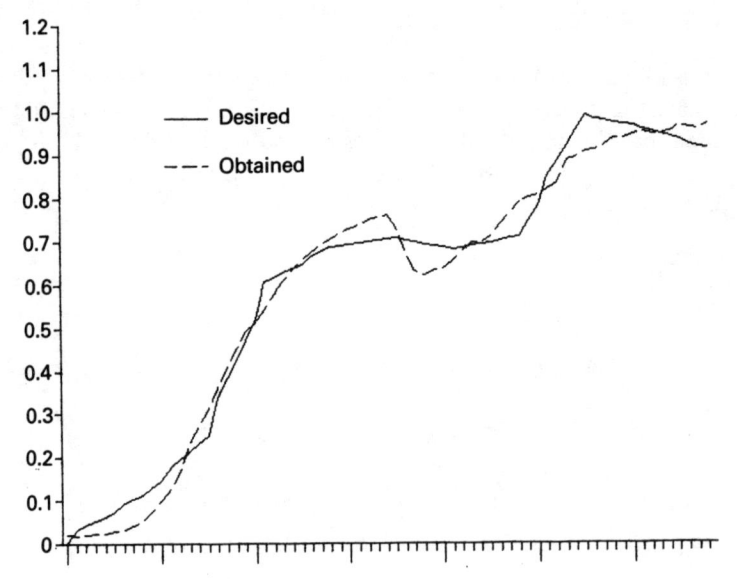

Fig. 5.14 Neural network estimations for phase II (learning set).

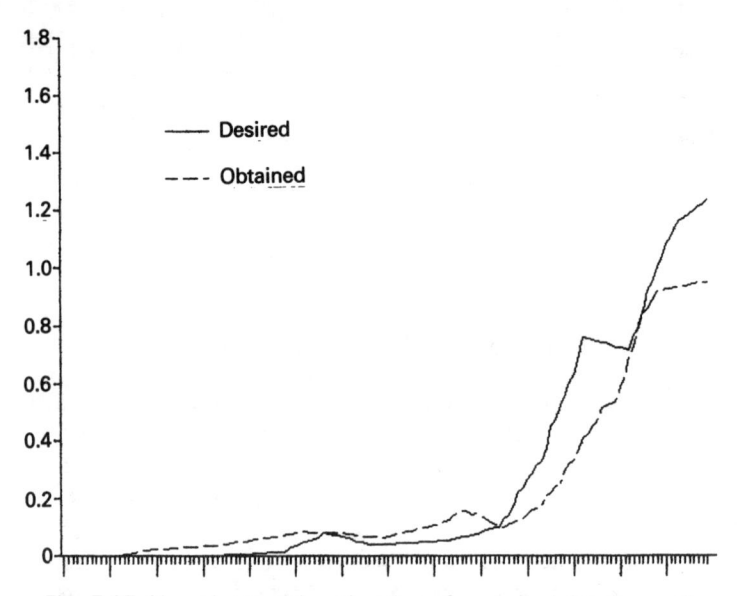

Fig. 5.15 Neural network estimations for phase I (testing set).

5.5 COMPARISON OF THE RESULTS

Two methods, namely identification and neural networks, have been compared using fermentation data. Despite the two methods differing in their approach, they both learn from a set of historic data and make predictions

Comparison of the results 69

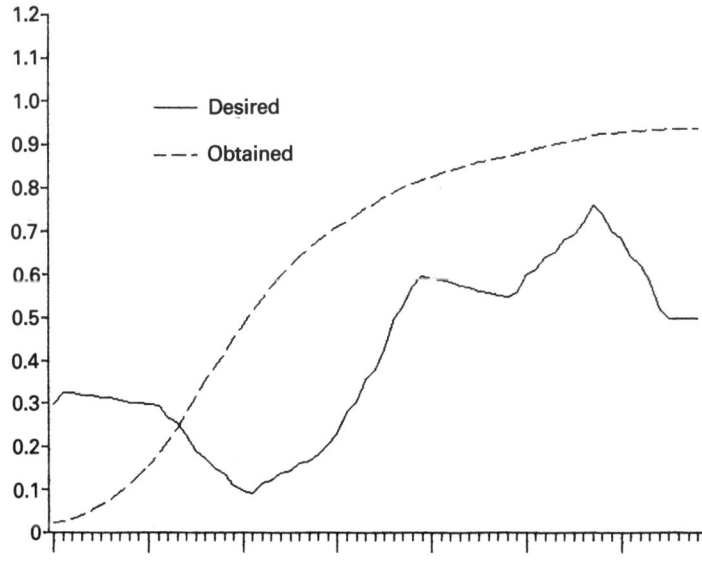

Fig. 5.16 Neural network estimations for phase II (testing set).

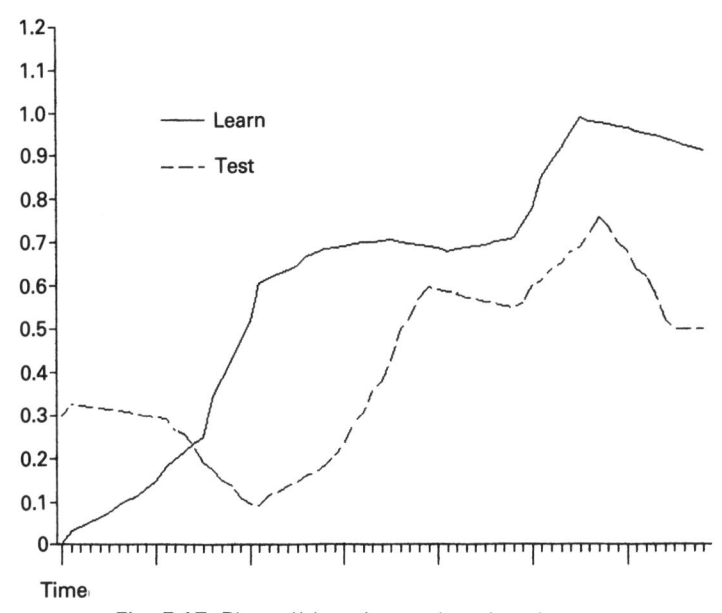

Fig. 5.17 Phase II learning and testing data sets.

on unseen new data and therefore a comparison is feasible. One can consider various comparison factors, for example speed of learning, in order to decide which method should be applied, but here the error is used as the primary criterion in measuring performance. By definition an error is the difference

70 State variables of a fermentation process

between the desired and the obtained class and since we consider all errors of equal importance the error is summarized over all examples. For the purpose of comparison all errors were normalized.

Considering Table 5.9 it seems that the neural network performance exceeded the identification performance during learning. The respective performances were analogous for phase I testing and only for phase II testing did the identification achieve a significantly better performance. Figures 5.18 to 5.21 illustrate the error curves. By their use, one can more easily perceive the distribution of the errors, in particular that the bulk of the identification technique errors are in the middle of each phase (learning) and that errors during testing have no common structure and seem erratic.

Table 5.9 Comparison of normalized errors

Technique	Learning		Testing	
	phase I	phase II	phase I	phase II
Model	24.93	42.57	28.42	32.98
Kalman	28.23	42.95	29.92	33.02
New model	25.17	42.00	28.75	31.67
Neural network	16.57	31.10	28.63	41.02

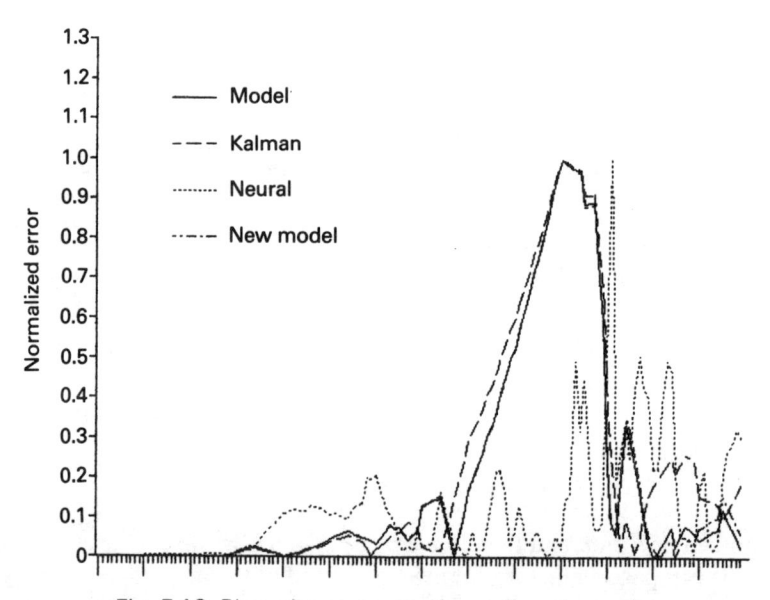

Fig. 5.18 Phase I error comparisons (learning set).

Comparison of the results 71

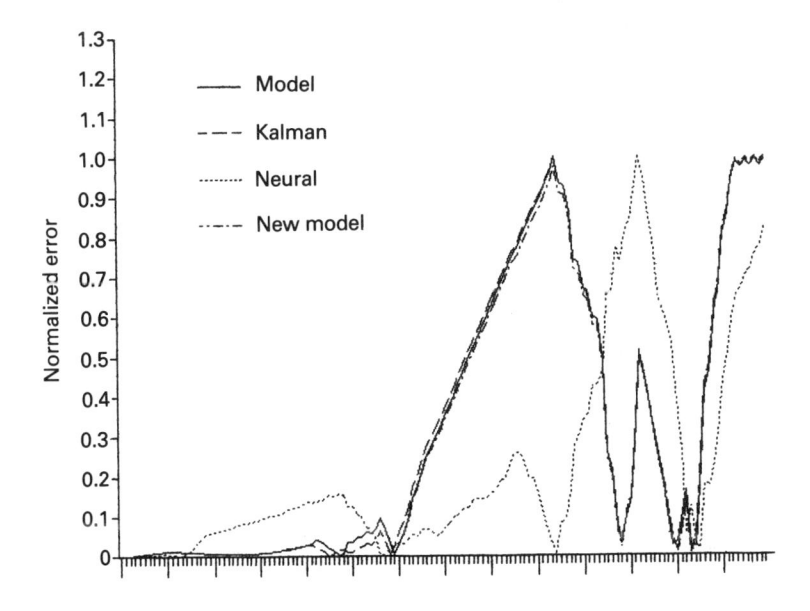

Fig. 5.19 Phase II error comparisons (learning set).

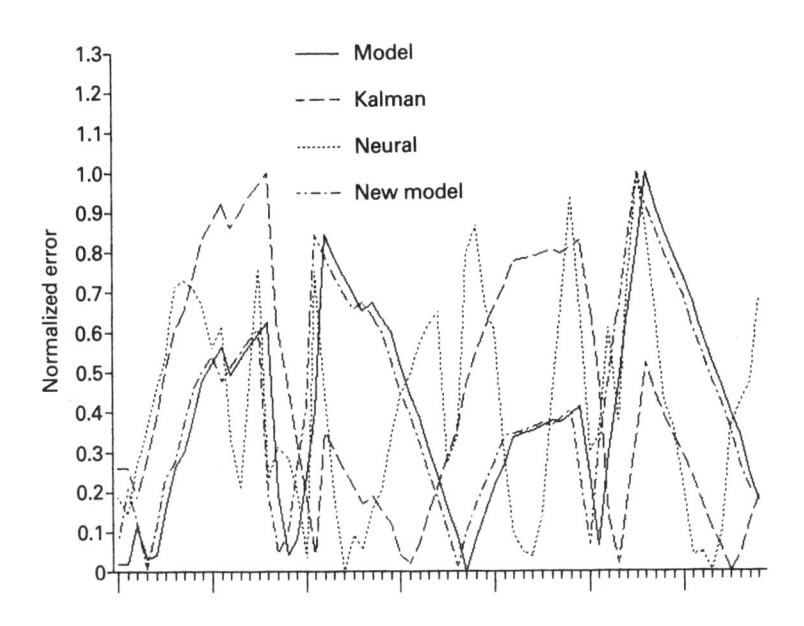

Fig. 5.20 Phase I error comparisons (testing set).

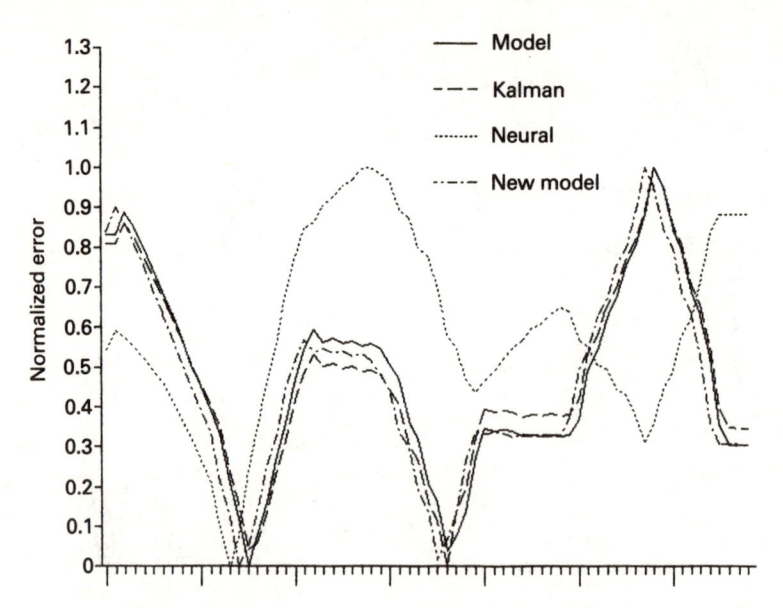

Fig. 5.21 Phase II error comparisons (testing set).

5.6 CONCLUSIONS

The comparison indicated that with an appropriate learning set the neural network will outperform the identification approach. This is mainly due to the ability of neural networks to handle the non-linearities of the data set. Furthermore it was demonstrated that a reduced connected network follows a similar learning profile to a fully connected network. Therefore one can pick the network with fewer hidden nodes and/or connections since it is expected that the simpler model will generalize better. On the other hand the use of an identification technique within an overall control strategy is well defined whereas the integration of neural network models in control architectures needs to be further investigated.

REFERENCES

1. Mirzai, A.R., Dixon, K., Hinge, R.D. and Leigh, J.R. (1991) Approaches to the modelling of biochemical processes. *IEE International Control Conference.* Edinburgh, UK.
2. Jalel, N.A., Hass, V., Mirzai, A.R., Munack, A. and Leigh, J.R. (1992) Modeling the Cyathus Striatus fermentation process: comparison of three methods. Keystone, Colorado, USA.
3. Hass, V. and Munack, A. (1990) Experimental design of fermentations for model identification. *American Control Conference*, San Diego, USA.
4. Ljung, L. (1987) *System Identification-Theory for the User* Prentice-Hall, Englewood Cliffs, NJ.

5. Leigh, J.R. (1985) *Applied Digital Control.* Prentice-Hall, London, UK.
6. Lippman, R. (1987) An introduction to computing with neural nets. *IEEE ASSP Magazine*, **4**, (2), 4–22.
7. Rumelhart, D.E., Hinton, G.E. and Williams, R.J. (1986) Learning representations by back-propagating errors. *Nature*, **323**, 533–6.

A practical application of neural modelling and predictive control

J.T. Evans, J.B. Gomm, D. Williams, P.J.G. Lisboa and Q.S. To

6.1 ABSTRACT

This chapter describes a method of obtaining an accurate neural network model of a non-linear system and its use in a practical neural control scheme. A technique of coding the data presented to a neural network (called spread encoding) is described and it is shown that using this technique greatly improves the accuracy of the neural network model compared to the conventional method of applying a single data value to a single input node of the network. Simulation results are presented and then results are shown of modelling a real laboratory scale process. The resulting neural network model is then incorporated into a neural predictive control strategy and used to predict future process outputs, which are in turn used to calculate process control input values so as to make the process behave in the same way as a specified reference model.

6.2 INTRODUCTION

Recent years have seen a significant increase in interest in the application of artificial neural networks to the modelling and control of systems. This interest is mainly due to the learning capabilities of neural networks, and the fact that multi-layer, feedforward networks can approximate any non-linear function with arbitrary accuracy [1]. This chapter describes the application of the multi-layer perceptron (MLP) neural network, trained using standard back-error propagation, to obtain a representative model of a physical non-linear process over a wide operational region. Subsequent implementation of the neural network model in a predictive control strategy is described and results are presented of the scheme applied on-line to control a laboratory process.

The usefulness of the neural network model in a predictive control strategy depends strongly on the ability of the model to predict accurately multiple time steps ahead. A NARMAX-type structure [2, 3] for the neural network model is used in these studies, within which lagged process input and output data are used to predict future process outputs. With this type of model the method by which data is represented to the network greatly influences the model prediction accuracy because future predictions rely on previous network outputs that are fed back to the network inputs. Hence, prediction errors can accumulate resulting in poor prediction accuracy. One approach to improve the network accuracy is to include dynamics within the neural network structure as described in [4]. An alternative approach to obtain an accurate model of the system is presented here that utilizes a new data conditioning method whereby each process data value presented to the neural network is distributed over a number of input nodes, rather than just one as is usually the case. The prediction accuracy of a neural network model trained with this method is compared to that of a network trained with the conventional data conditioning method of applying the data for each network input to a single input node [2, 4–6].

When an accurate model of the system under consideration is obtained, it can be used in a variety of control strategies. Some forms of control strategies that have utilized neural networks include the following.

- Internal model control [3]. Here a radial basis function neural network was used to model a non-linear system and it was found that, to achieve a good level of accuracy, a network consisting of a very large number of processing nodes was required.
- Another possible approach of using neural networks in a control strategy is to use direct inverse control as described in [7]. The disadvantage of this method is that the control structure relies heavily on the ability to obtain an inverse model of the system, which may not always be possible.
- A third method is that of model reference control where the performance of the system under closed loop is specified by a stable reference model. Implementation of this control structure is described extensively in [8].

The control strategy implemented in this chapter is that of predictive control which directly utilizes the neural model to predict future process outputs and appropriate input control signals are computed to minimize a specified cost function.

6.3 DATA CONDITIONING

Two forms of data conditioning have been studied in this work and a comparison between the neural network's performance and relative merits in both cases is made in later sections. The standard data conditioning technique is to linearly map the data, between a specified lower and upper bound, for each network input and output node and apply the resulting values to single, individual network nodes [2, 4–6], the effective operational range of the network input and output nodes frequently being between 0.1 and 0.9. This approach will be referred to as single node data mapping, SNDM.

A new method of conditioning the data is proposed whereby each data value is spread over a number of nodes using a sliding Gaussian distribution as shown in Fig. 6.1 (referred to as spread encoding, SE). The range of the data is evenly discretized across the nodes such that each node has a value in the data range assigned to it. The coded data value is represented at the centre of the Gaussian excitatory pattern and is retrieved as a weighted sum of the excitations and data values of each node. The translation equations used for the spread encoding technique are described in Chapter 2 of this book and in [9]. In the results presented, each data value was spread over six nodes.

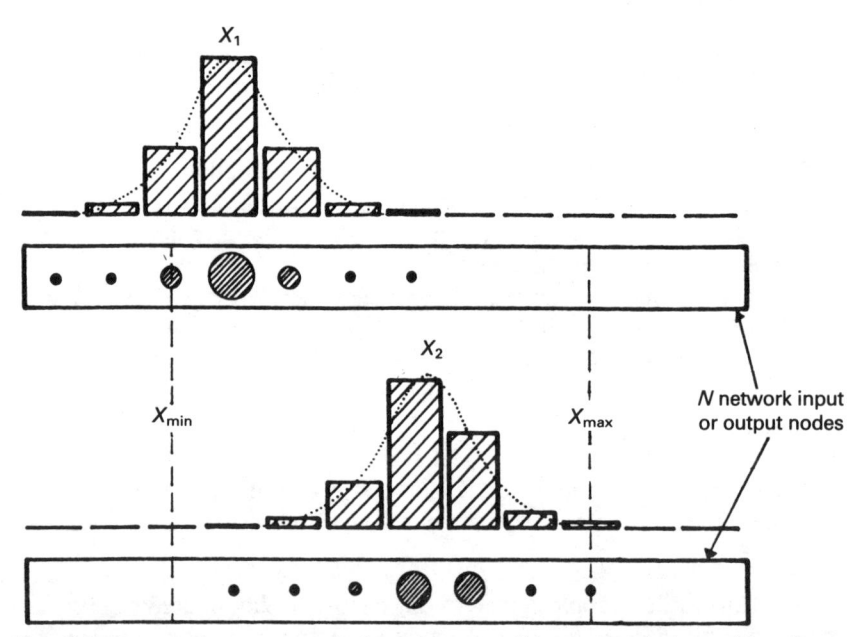

Fig. 6.1 Proposed spread encoding data conditioning technique. The figure illustrates the coding of two data values, x_1 and x_2 ($x_2 > x_1$), in the range (x_{min}, x_{max}) to N network nodes.

The SE technique has been investigated since, in the area of measurement and process control, the physical variables of interest usually span a wide range of analogue values. If the range of these values is compressed to a single node, as in the case of SNDM, the amount of activity in relevant modes of operation may span only a small fraction of the full range of the processing node. The spread encoding technique overcomes this disadvantage by distributing the information over a number of the network nodes. This technique was also considered since many biological systems work on the principle of spreading the information that they receive, e.g. the retina of the eye.

Results are presented that show that the spread encoding technique greatly improves the accuracy of the neural network to model a process, compared to that of SNDM. One reason for this increase in accuracy could be that when using the spread encoding technique the problem is turned into one of a pattern recognition problem, which is well suited to neural network technology. The spread encoding technique increases the dimensionality of the network weight space, and this may at first seem a serious disadvantage. However, studies indicate that when using the SE technique the neural networks require fewer complete passes of the training data set and give a lower mean squared error than networks trained using the alternative SNDM method, as illustrated in Fig. 6.4 below.

6.4 THE PROCESS

The system considered in this paper is a dual-tank, non-interacting liquid-level system as shown in Fig. 6.2. Both the tanks are of the same size and capacity, the height being 1.25 m and the capacity of each tank being 20 litres. The process has inherent non-linearities and, to be more representative of systems encountered in practice, an unmeasured state in the height of liquid in the top tank. The non-linearity present in this system is due to the square root relationship between the tank outflows and the height of liquid in each tank. The laboratory process uses standard industrial equipment (Fig. 6.2). A pressure measurement, to measure liquid-level height, is taken from the bottom of tank 2. The signal is converted to a suitable voltage (in the range 0–5 V) for A/D conversion via a standard pressure to current converter (3–15 psi, 4–20 mA). The computer-controlled driving output (the control input) is a digital number between 100 and 200, which is converted to a pressure to drive the valve through its full operational range. The neural network and control software for the process were written in the Quick Basic programming language and implemented on an IBM PS/2 model 30 computer.

Initially a mathematical model of the liquid-level system was developed with the same characteristics as the laboratory system. The mathematical

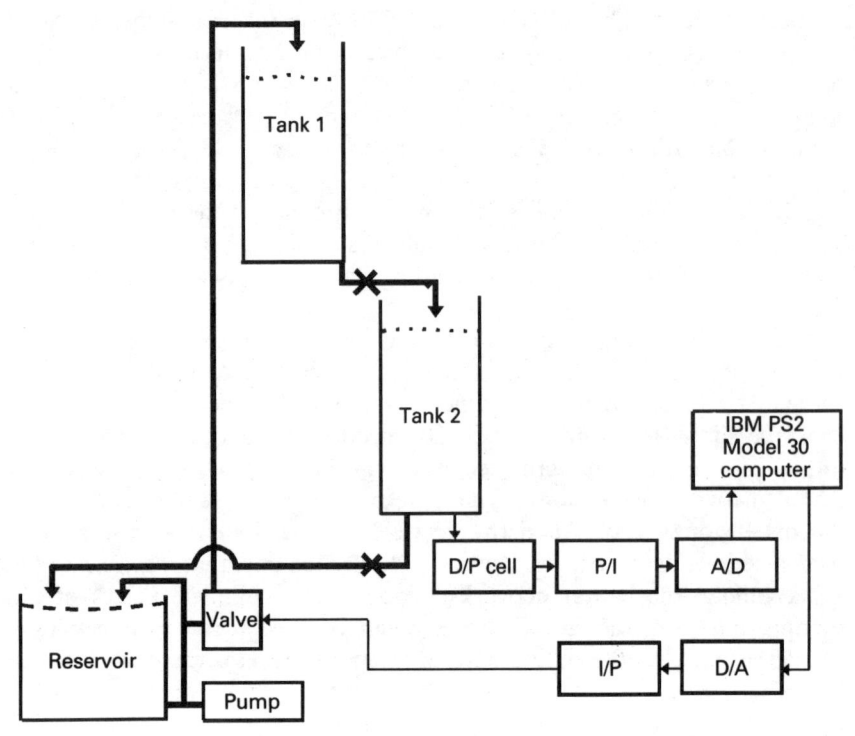

Fig. 6.2 Liquid-level laboratory process.

model of the liquid-level process was represented by the following well-known, non-linear differential equations:

$$\dot{h}_1 = \frac{1}{C_1}\left(K_v u - \frac{\sqrt{h_1}}{R_1}\right) \qquad (6.1)$$

$$\dot{h}_2 = \frac{1}{C_2}\left(\frac{\sqrt{h_1}}{R_1} - \frac{\sqrt{h_2}}{R_2}\right) \qquad (6.2)$$

where h_1 and h_2 are the liquid levels in tanks 1 and 2 respectively, u and h_2 are the process input and output, K_v is the valve gain, C_1, C_2, are the cross-sectional areas of tanks 1 and 2 respectively and R_1, R_2 are the outflow pipe restrictances of tank 1 and tank 2. The mathematical process model was simulated, using a standard continuous simulation package, in order to obtain data for training and validating the neural networks. Subsequently, the same approach was applied to the real process.

6.5 DEVELOPING A NEURAL NETWORK MODEL OF THE SYSTEM

An artificial neural network (ANN) model of the dual-tank, liquid-level system was initially developed prior to developing a control algorithm. In the development of this model, there are a number of issues that need consideration, such as the network topology, network training and validation, and these are discussed in the following subsections.

6.5.1 NEURAL NETWORK TOPOLOGY

It has been assumed that the input–output dynamics of the liquid-level process can be characterized by the general NARX (Non-linear, Auto-Regressive, eXogenous) model defined by:

$$y(t) = F(y(t-1), \ldots, y(t-n_a), u(t-k), \ldots, u(t-k-n_b)) + e(t) \qquad (6.3)$$

where F is some unknown non-linear function, y and u are the process outputs and inputs respectively, e is a Gaussian distributed white noise sequence, k is the process deadtime, n_a and n_b are the number of past output and input data used in the above non-linear difference equation. The approach taken was for the neural network model to have the same number of present and past values in its structure as would be taken in NARX input–output modelling. The results are presented for the ANN model structure: process deadtime, $k=1$; $n_a=2$ and $n_b=1$. Hence, two past inputs and two past outputs of the process are used as inputs to the neural network. These values result in the SNDM neural network having four input nodes and one output node. For the SE network, where each data value was mapped over six nodes, the topology consisted of 24 input nodes and six output nodes. In both cases just one hidden layer was used for each network structure. The number of nodes used in the hidden layer was determined empirically [9]. For the SNDM network eight hidden nodes were used and six nodes in the SE network. Each of the hidden and output nodes utilized a sigmoidal activation function. The next step, in the development of the neural network model, was to train the networks to be able to represent the dynamics of the dual-tank, liquid-level system.

6.5.2 NETWORK TRAINING

The neural networks are trained using the back-propagation algorithm, which is a form of the well-known, gradient descent algorithm. There are two algorithm parameters, namely, the gain and momentum which have to be selected to control the iterative process. The values of the gain and momentum were initially set at 0.9 and 0.6 respectively, and were manually reduced as the network training proceeded. If the gain term is initially small,

80 A practical application of neural modelling

then training the network can take a long time. However, if the initial value of the gain is not reduced during the training period the network may not reach its global minimum, since the large step size causes the network to oscillate around its global minimum

Both the neural networks (SNDM and SE) were trained as one-step-ahead predictors (Fig. 6.3(a)) because this is a more stable training configuration with the back-propagation algorithm than the model configuration [5]. It has been found, in this research, that if a neural network is overtrained, its generalization properties deteriorate when structured as a model of the process (Fig. 6.3(b)). On the other hand, if not trained long enough the generalization properties and accuracy of a network are poor. The approach used to select the optimum amount of training is as described in [9].

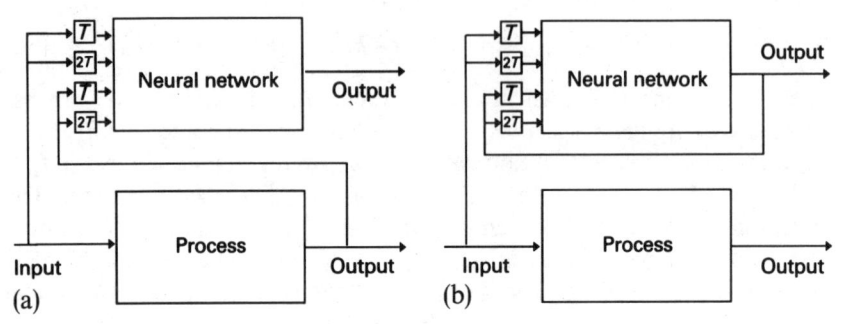

Fig. 6.3 Neural network training and recall configurations for process modelling: (a) predictor structure; (b) model structure.

Two different input–output data sets (D1 and D2), each containing 1000 pairs of input–output values, were generated from simulations of the mathematical model describing the dual-tank, liquid-level process with a random amplitude signal applied to the process input. One of the data sets (D1) is used to train the neural network in the one-step-ahead predictor configuration and the other (D2) is used to test the network's generalization properties in the model configuration at various stages of the training. Figure 6.4 shows the mean squared error of the network outputs for D2 (after decoding and scaling back to the original data range) against the number of complete passes of D1 for both the SNDM and the SE neural networks. The optimum amount of training is indicated by the lowest point in the graphs. Beyond this point it can be seen that the generalization properties as a model degrade for both the SE and SNDM networks. Figure 6.4 also shows that the SE network requires a smaller number of complete passes (300), of the training data, than that required for the SNDM network (700). Furthermore, the mean squared error for the SE network is significantly less than that of the SNDM network.

Developing a neural network model 81

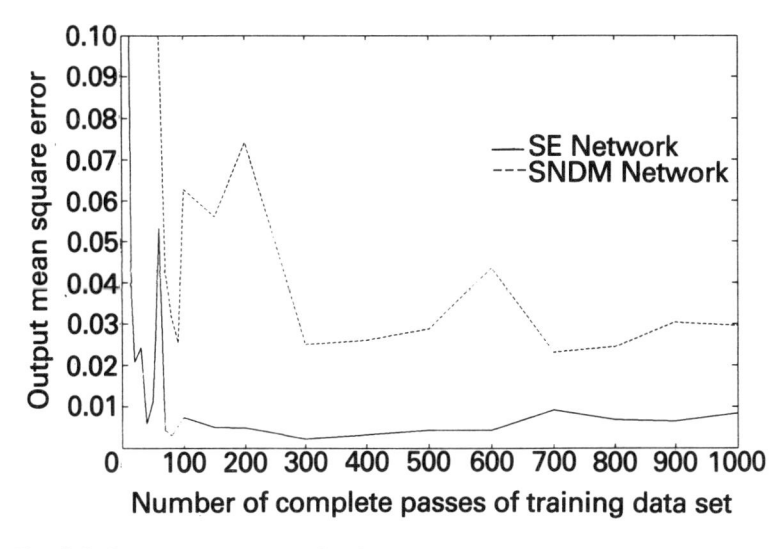

Fig. 6.4 Generalization tests for SNDM and SE networks during training.

6.5.3 MODEL VALIDATION

The neural networks were validated as both a one-step-ahead predictor and also as a model of the process. Figure 6.3(b) shows that when structured as a model, the network inputs consist of the past input data from the process and the past delayed output data from the neural network. Using the model configuration allows the network to be used independently of the process which is a desirable property. The two trained neural networks were validated in both structures on a number of test signals that had not been used in training. These included a random amplitude signal, a pseudo-random binary signal (PRBS) and a set of steps.

Figures 6.5 and 6.6 show the simulation results when recalling both the SE and SNDM networks as one-step-ahead predictors on a set of step responses. The set of steps presents a severe test for both the networks, since it spans a wide region of operation and indicates whether the neural networks have learned the required steady states. It can be seen from Figs 6.5 and 6.6 that both the networks perform adequately, the SE network having a slightly less mean square error than the SNDM network. However, when testing the two neural networks on the set of steps in the model configuration the SE network (Fig. 6.8) performs significantly better than that of the SNDM network in Fig. 6.7. This performance was repeated when the networks were tested with the PRBS and random amplitude signals and is shown in [10].

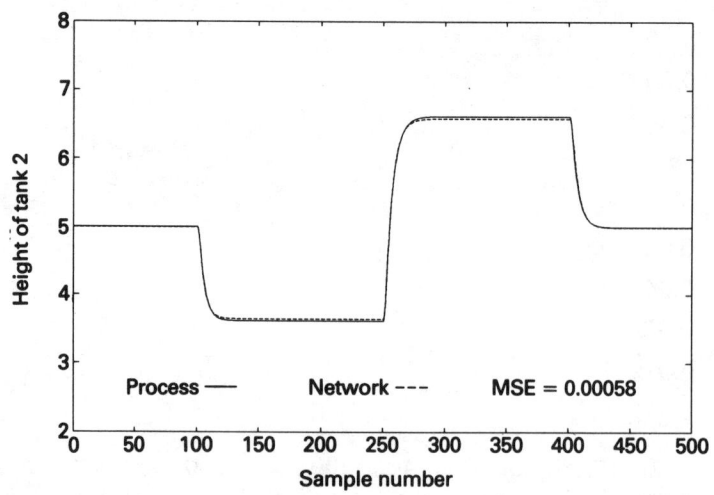

Fig. 6.5 SNDM network tested on a series of step inputs as a one-step-ahead predictor.

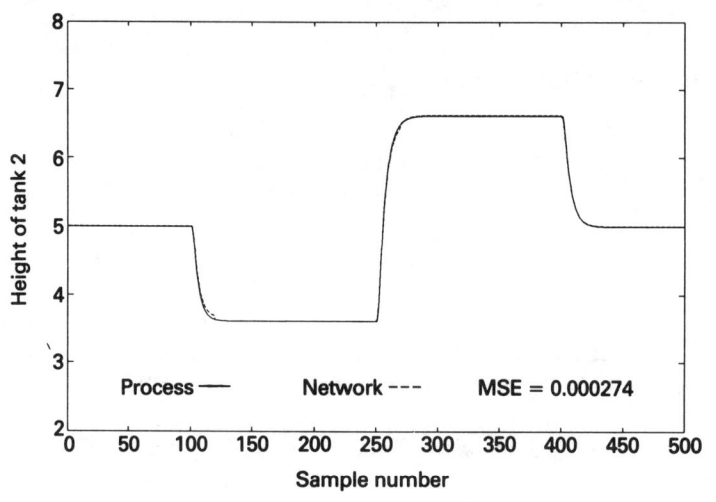

Fig. 6.6 SE network tested on a series of step inputs as a one-step-ahead predictor.

6.5.4 REAL PROCESS RESULTS

The simulation results described above indicate that spread encoding the data presented to the neural network gives a significant improvement in accuracy when using the neural network in a model configuration, and is slightly better when in the one-step-ahead predictor configuration. Hence, the laboratory process was only modelled using the spread encoding technique. Figure 6.9 compares the responses of the real process and that of

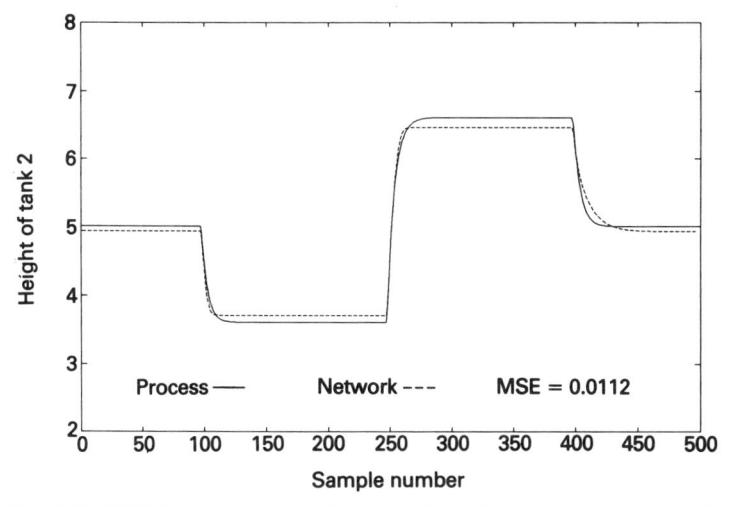

Fig. 6.7 SNDM network tested on a series of step inputs as a model.

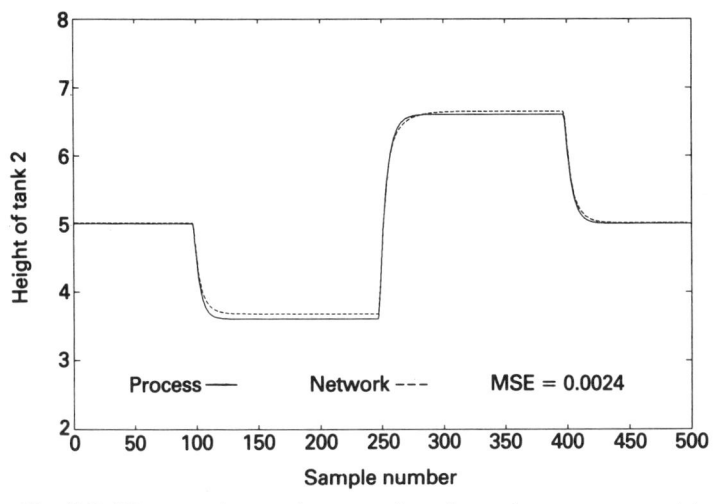

Fig. 6.8 SE network tested on a series of step inputs as a model.

the trained neural network when tested in a model configuration on a PRBS. The neural network accurately simulates the response of the process. It can be seen that on the second sequence of the PRBS signal, the height of liquid in tank 2 does not reach the same level as it did on sequence 1, although it does reach the level on the third sequence. This is indicative of time variations present in real processes and illustrates the differences when using a simulation model and data from a real process. The neural network model of the laboratory process was then recalled on a random amplitude signal as shown in Fig. 6.10, and again an acceptable result is obtained. The

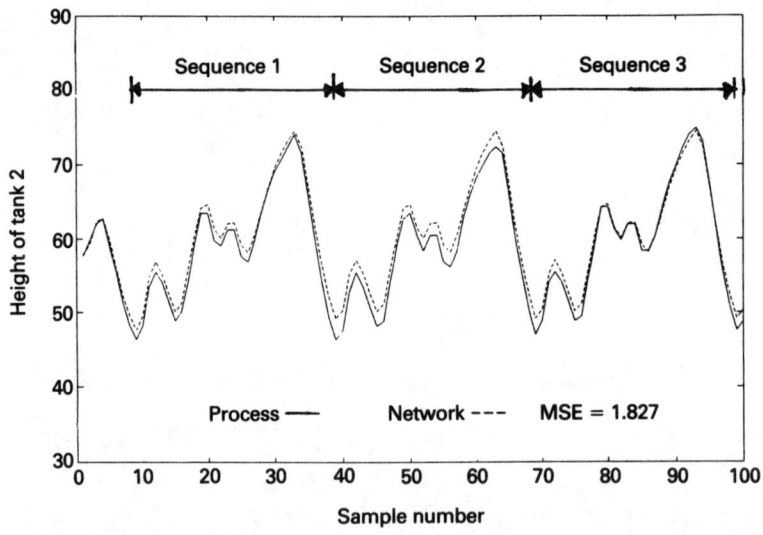

Fig. 6.9 Responses of SE network model and laboratory process to a PRBS.

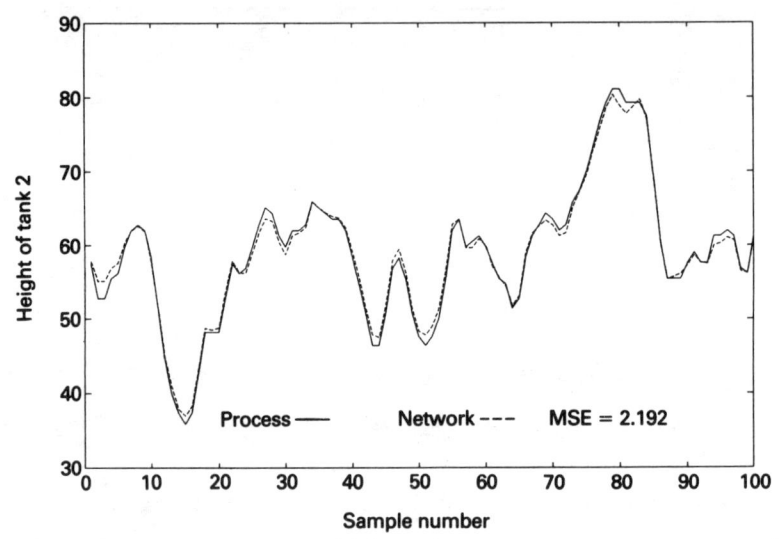

Fig. 6.10 SE network model and laboratory process responses to a random amplitude signal.

neural network model was also successfully validated as a one-step-ahead predictor on the above two signals.

6.6 NEURAL PREDICTIVE CONTROL: A PRELIMINARY STUDY

Having developed a suitable neural network model of the real system, it can be utilized in a control structure. The control structure adopted is illustrated in Fig. 6.11 and is that of neural predictive control. The neural network model is used to predict future process outputs, and the control signals are chosen so as to minimize a suitable cost function in terms of future output deviations from the required set point values. The control signals are specified by minimizing the following cost function:

$$J = \sum_{k=1}^{N} (y_{ref}(t+k) - y_{nn}(t+k))^2 + \sum_{k=1}^{N} \lambda(u(t+k-1) - u(t+k-2))^2 \quad (6.4)$$

where y_{ref} and y_{nn} are the outputs of the reference model and neural network respectively, λ is a weighting factor in the interval (0, 1), u is the controller

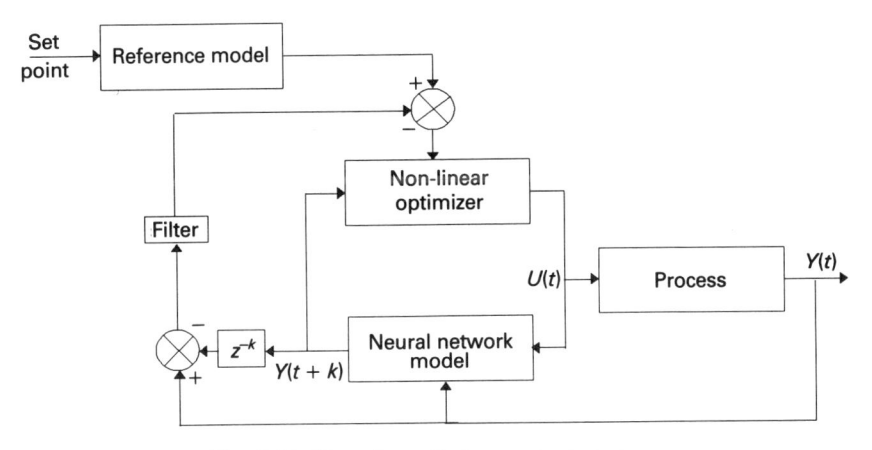

Fig. 6.11 Neural predictive control strategy.

output and N is the prediction horizon. In the results presented only one-step-ahead prediction was considered, hence, $N=1$ in equation (6.4). A reference model of the form:

$$G(s) = \frac{\omega_n^2}{s^2 + 2c\omega_n s + \omega_n^2} \quad (6.5)$$

was used to specify how the liquid-level process was required to behave. For this study the damping ratio, c, was chosen as 0.7 and ω_n was calculated to obtain a 2% settling time of 7 minutes.

6.6.1 ON-LINE RESULTS

Figure 6.12 shows the required reference model output and the height of the second tank in the system for λ, in equation (6.4), equal to zero, i.e. no constraint on the input value to the valve. It is clearly seen that the process tracks the reference model as required. Figure 6.13 shows the corresponding control input which shows that with λ set to zero the control input is not a

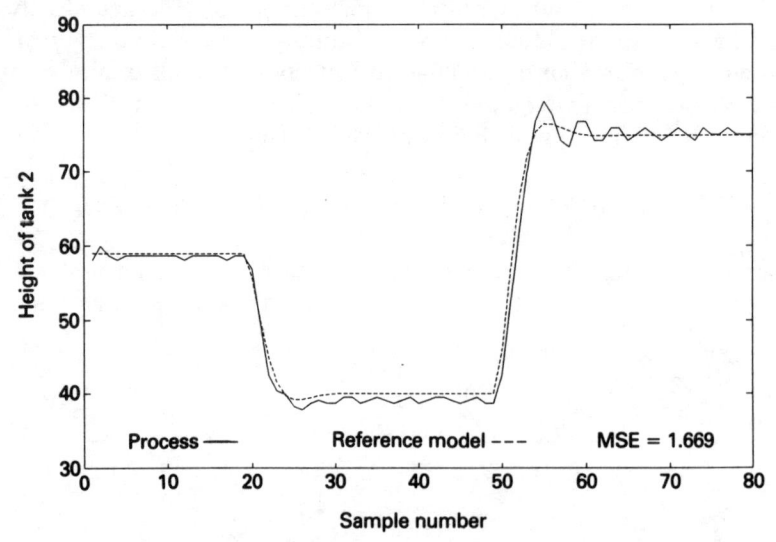

86 A practical application of neural modelling

Fig. 6.12 On-line predictive control results for $\lambda = 0$ in equation (6.4).

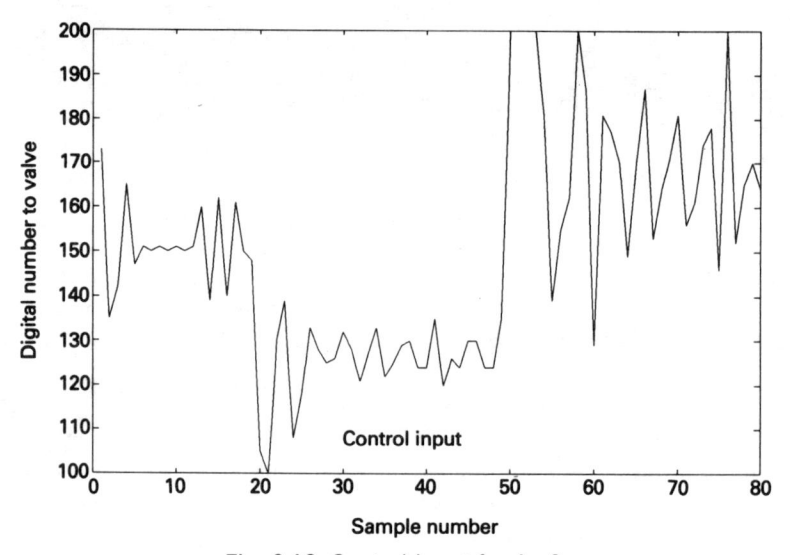

Fig. 6.13 Control input for $\lambda = 0$.

desirable one since the valve in the system is continually subjected to excessive movement. The effect of increasing the weighting factor, λ, to 1 can be seen in Figs 6.14 and 6.15. The control valve movement is much smoother but at the expected expense of a poorer control performance.

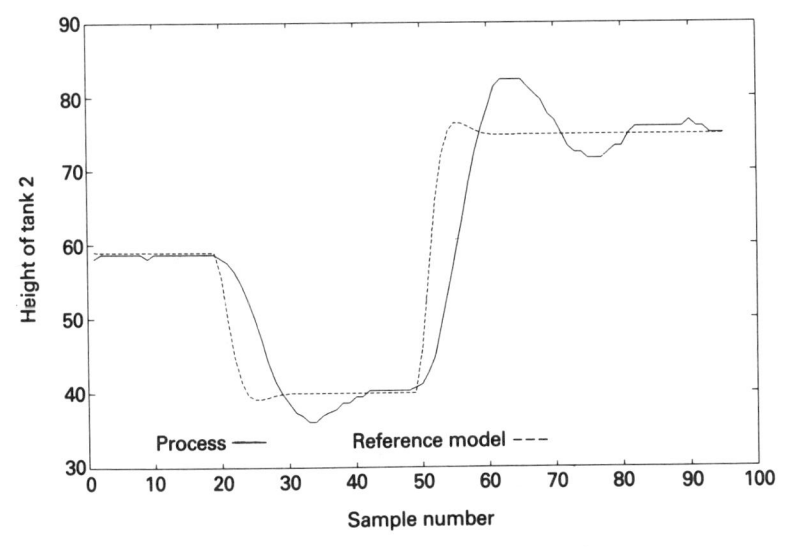

Fig. 6.14 On-line predictive control results for $\lambda=1$.

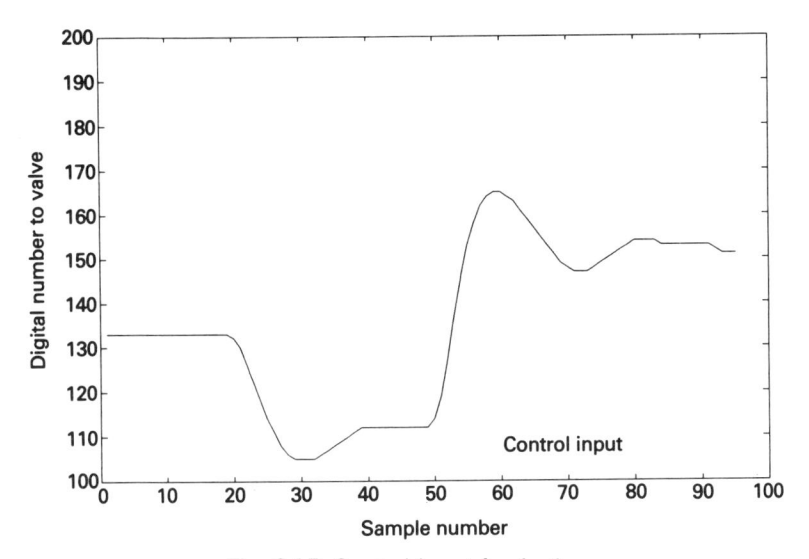

Fig. 6.15 Control input for $\lambda=1$

6.7 CONCLUSIONS AND FURTHER WORK

It has been shown that it is feasible to use artificial neural networks to model non-linear dynamic processes, and using the proposed spread encoding data conditioning technique results in improved accuracy when the neural network is required to act as a process model. The neural network representation of the real process has been successfully incorporated into a predictive control strategy, and on-line results were presented which demonstrate that stable and accurate control is possible.

Work is in progress to investigate multi-step-ahead predictive control, as this should attenuate the excessive movement of the valve which occurred for one-step-ahead prediction when no constraint on the control input was present. Also other control strategies implementing neural networks are being investigated and a comparison of the neural network methods and conventional PID control is to be made. The spread encoding of the process data is also to be implemented on a more complex modelling and control problem.

ACKNOWLEDGEMENTS

The authors would like to acknowledge the Science and Engineering Research Council for financial support of J.T. Evans and Q.S. To, and also Unilever Research Laboratories for CASE support of Q.S. To.

REFERENCES

1. Hornik, K., Stinchcombe, M. and White, H. (1989) Multistage feed-forward networks are universal approximators. *Neural Networks*, **2**, 359–66.
2. Chen, S., Billings, S.A. and Grant, P.M. (1990) Non-linear system identification using neural networks. *Int. J. Control*, **51**(6), 1191–1214.
3. Hunt, K.J. and Sbarbaro, D. (1991) Neural networks for non-linear internal model control. *IEE Proc. D.*, **138**(5), 431–8.
4. Willis, M.J., Di Massimo, C., Montague, G.A., Tham, M.T. and Morris, A.J. (1991) Artificial neural networks in process engineering. *IEE Proc. D.*, **138**(3), 256–66.
5. Narendra, K.S. and Parthasarathy, K. (1990) Identification and control of dynamical systems using neural networks. *IEEE Trans. Neural Networks*, **1**(1), 4–27.
6. Bhat, N.V., Minderman Jr, P.A., McAvoy, T.J. and Sun Wang, N. (1990) Modelling chemical process systems via neural computation. *IEEE Control Systems Magazine*, **April**, 24–9.
7. Barto, A.G. (1990) Connectionist learning for control: an overview. In *Neural Networks for Control* (ed. W. Thomas Miller III, R.S. Sutton and P.J. Werbos). M.I.T. Press, Cambridge, Mass., 5–58.
8. Narendra, K.S. and Parthasarathy, K. (1989) Neural networks and dynamical systems. Part III: Control. Internal Report No. 8909, Yale University.
9. Gomm, J.B., Lisboa, P.J.G., Williams, D., Evans, J.T. and To, Q.S. (1992) Neural networks for accurate modelling of non-linear process dynamics. Submitted to *IEE Proc. D.*, July 1992.
10. Evans, J.T., Gomm, J.B., Williams, D., Lisboa, P.J.G. and To, Q.S. (1992) Non-linear process modelling with artificial neural networks. *Proc. 24th SCS Summer Computer Simulation Conference*, Reno, USA, 27–30 July, pp. 637–41.

A label-driven CMAC intelligent control strategy

7

A.J. Lawrence and C.J. Harris

7.1 ABSTRACT

An *a priori* unknown SISO linear plant, subject to parametric variations, is posed as a control problem for the development of a cerebellar model articulation controller (CMAC) based adaptive control strategy. The plant output is characterized by labels (e.g. rise time, overshoot, steady state error) which serve the purpose of data compression whilst extracting salient features for control. Desired labels are generated from a reference model. Two CMAC modules are required for plant modelling and controller tuning. On-line tuning is performed by issuing a step demand to both the CMAC and reference models, the outputs of which are used to generate a label error vector. A CMAC tuning module maps the label error to a controller gain change and the controller is updated. The CMAC algorithm is capable of learning a wide class of non-linear mappings from training data and hence is not explicitly programmed.

7.2 INTRODUCTION

This work is part of a project which aims to develop an intelligent strategy for the control of aeronautical gas turbine engines. An adaptive control strategy is described and some preliminary results presented.

The cerebellar model articulation controller (CMAC) was developed for solving robot arm articulation problems in which the many degrees of freedom made conventional approaches difficult [1, 2, 3]. CMAC has been successfully applied to robot control problems [4] and pattern recognition, signal processing and adaptive control [5]. CMAC belongs to a class of neural networks (NNs) which consist of a fixed non-linear input layer coupled to an adjustable linear output layer. Other types of associative memory networks include radial basis functions (RBFs), *B*-splines and fuzzy logic (FL). Some of the features which make this class of NNs attractive for adaptive control are their capability for learning non-linear mappings, fast learning convergence and globally stable incremental learning.

The objective of an adaptive control scheme is to adapt the parameters of the controller in response to changes in the dynamics of the plant in order to achieve a desired level of control. Often this is achieved by assuming a model structure for the plant and identifying its parameters on line. The adaption of the controller parameters is then based on the identified model. A distinction is made between adaptive control as described above and conventional gain scheduling which splines controllers, designed off line, for some operating point and thus is invariant under plant dynamic changes. Clearly an adaptive control strategy is not required if the dynamics of the plant are invariant. The ultimate aim of this work is to apply an adaptive control strategy to an aeronautical gas turbine engine which, among other characteristics, exhibits non-linear and time-varying dynamics.

The work presented in this chapter is reminiscent of self-tuning strategies such as the well-known Ziegler–Nichols tuning rules. A small set of empirical measures taken from the response is required to determine the controller parameters. In this way a restricted class of processes may be characterized and controlled in a simple but effective manner. It is hoped that this methodology will widen this class of processes by using user-defined labels for process characterization and by being independent of controller structure.

In order to develop an adaptive control strategy a control task is required that does not mask fundamental issues; for this purpose we have chosen a SISO linear plant in which time-varying dynamics are simulated by parametric changes.

7.3 THE CMAC ALGORITHM

The CMAC algorithm belongs to a class of neural networks which consist of an adjustable combination of fixed basis functions and a linear update rule. This approach allows learning and stability conditions to be readily derived [6]. The action of this general class is depicted in Fig. 7.1 in diagrammatic form and is described by (7.1).

$$h_j = \phi(\|x - \gamma_j\|)$$

$$\hat{y}_k = \sum_{j=1}^{M} w_{jk} h_j \qquad (7.1)$$

The CMAC algorithm 91

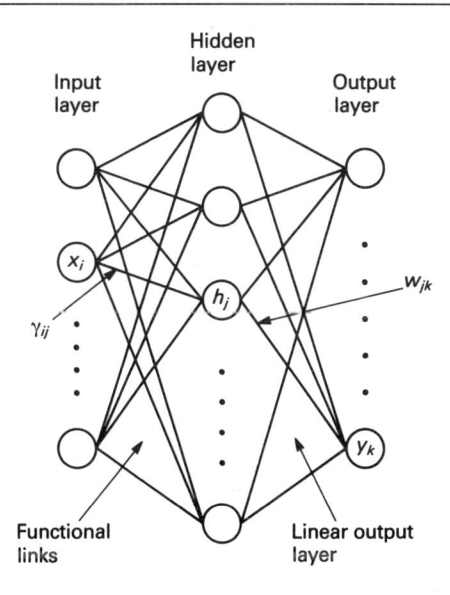

[Figure: A neural network diagram with Input layer (containing x_i), Hidden layer (containing h_j), and Output layer (containing y_k), with labels γ_{ij} for Functional links and w_{jk} for Linear output layer connections.]

Fig. 7.1 A linear combination of fixed basis functions.

$\phi = \phi(r)$ is a basis function, γ_j is the jth basis function centre and h_j is its height at the input x, w_{jk} is the contribution of the jth basis function to the kth component of the output ($\hat{y}_k$), and M is the total number of basis functions and hence is proportional to the memory requirements. The CMAC algorithm with piecewise constant basis functions is described by (7.2) for a normalized output.

$$\phi(r) = \begin{cases} 1, & r \leq \dfrac{\rho}{2} \\ 0, & r > \dfrac{\rho}{2} \end{cases} \qquad (7.2)$$

where

$$r = \| x' - \gamma_j \|_\infty$$

$$\hat{y}_k = \frac{\sum_{j=1}^M w_{ij} h_j}{\sum_{j=1}^M h_j}.$$

Note that x' represents x expressed in quantized coordinates. Due to the infinity norm and with an appropriate choice of basis function centres, (7.2) gives rise to ρ regular N-dimensional lattices (N = input dimension) in which each $N-d$ cube is of side ρ quantization intervals. The lattices are so arranged that any input x falls within exactly ρ $N-d$ cubes, one from each lattice. This observation allows the network output to be written as (7.3) in which * represents the selected weights:

$$\hat{y}_k = \frac{\sum_{j=1}^{\rho} w_{jk}^*}{\rho}. \tag{7.3}$$

The network is trained in a supervised manner. It is forced to emulate (x, y) pairs by applying the update rule given by (7.4) to each such pair. The training set or data consists of all such pairs and may be presented to the net several times.

$$w_{jk}^n = w_{jk}^{n-1} + \eta \varepsilon^n \qquad \varepsilon^n = y_k - \hat{y}_k^n \tag{7.4}$$

where $n=$iteration index, and $\eta=$learning rate.

The update rule (7.4) is a least mean squares algorithm and enables the network to find a weight vector which best matches the training data in a least squares sense. Note that it is the ρ active weights that are updated after each training pair. The accuracy of the solution is dependent on the learning rate, η, and the suitability of the basis functions for approximation of the desired function. For this work the CMAC algorithm was configured as follows. The univariate basis functions were chosen as piecewise linear 'hat' functions combined under the product rule to form the multivariate basis functions. Learning was carried out using the normalized least mean square update rule. The lattice placement scheme developed by Parks and Militzer was adopted [7]. A decaying learning rate was used, the decay parameter set to 50.0. These modifications to the Albus CMAC (and others) are described in [8]. The net description is summarized in Table 7.4 and includes the input space quantization (knot matrix) plus the total number of memory locations.

7.4 CONTROLLER DESIGN AND OPERATION

7.4.1 CONTROL SCHEME

The control scheme investigated in this chapter is given in Fig. 7.2, and is an extension of the work carried out by Fraser [9]. It may be divided into four modules,

1. an error driven controller and unity feedback loop;
2. a CMAC plant model;
3. a label error calculation module; and
4. a CMAC controller tuning module.

The CMAC plant model is continuously updated on line adapting to changes in the dynamic behaviour of the plant. Note that this includes sensor and actuator dynamics. For this purpose some knowledge of the

Controller design and operation

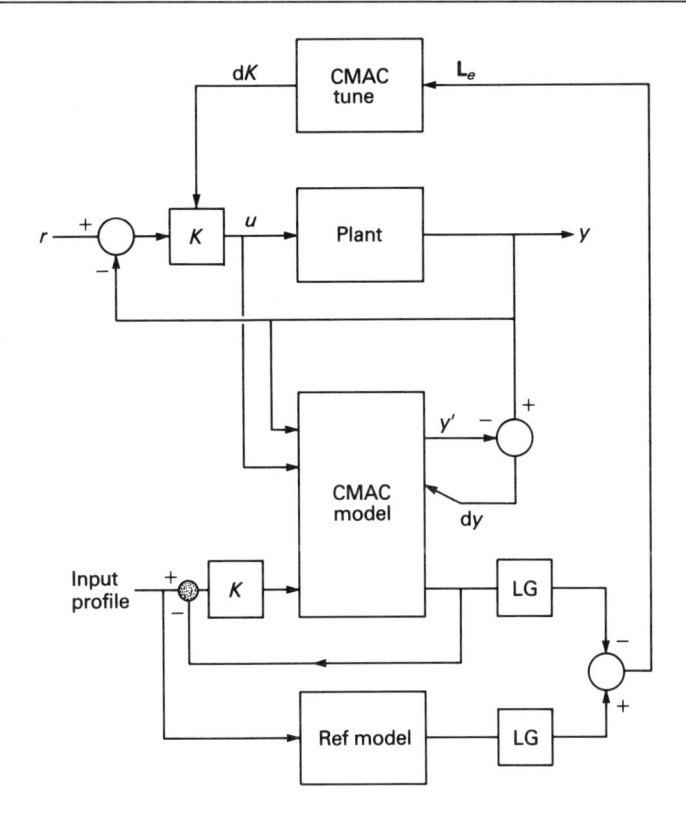

Fig. 7.2 Possible control system architecture. LG = label generation.

plant is required to estimate the model order, that is, values of r and a in the discrete, non-linear, plant model (7.5).

$$y_k = f(u_{k-r}, u_{k-r+1}, \ldots, u_k, y_{k-s}, y_{k-s+1}, \ldots, y_{k-1}). \tag{7.5}$$

A step change in demand is imposed on the closed loop plant model and the desired model. A label vector is generated for each of the responses and the label error vector is calculated according to (7.6) where pm and rm denote plant and reference models respectively.

$$L_e = L_{pm} - L_{rm}. \tag{7.6}$$

The label error is mapped to a gain change by the CMAC tuning module and the controller gains are updated. This process may be carried out continuously on line without command input restriction.

The label vectors fall into two categories; representation of process dynamics and static constraints. Prime dynamic characteristics of the process response are represented by time-invariant labels such as rise time, overshoot, steady state error, as described here, or novel labels specific to an application. Labels describing static constraints, such as maximum engine r.p.m., have not been investigated here but are expected to provide the control system with the ability to observe operating limits. The use of labels to describe the response provides significant data compression, e.g. a sampled response containing in excess of 100 data points may be satisfactorily described by several labels. Furthermore, the emphasis is placed on the salient response features for control.

If the labels chosen provide a good description of the important features of the transient response then the label errors, as defined by (7.6), provide a quantitative measure of control system performance relative to the chosen reference model labels. Some norm of a normalized and weighted label error vector would be required to provide a consistent performance factor. However by comparing Table 7.3 with Fig. 7.3 we see that small label errors correspond to good system performance. This is discussed in more detail in section 7.5.

The above approach differs from that presented by Kumar and Guez, [10] in which label extraction provides a neural net with a means of plant model identification. Control law synthesis is then a function of the identified model parameters. In the approach presented here, a model is used for training data generation but its structure is unrestricted. Hence a non-parametric neural network plant model may be used for this purpose.

7.4.2 THE CMAC TUNING MODULE

In the control scheme described the tuning module performs the mapping in (7.7).

$$L_e \to \Delta K. \qquad (7.7)$$

The CMAC is trained in a supervisory mode, hence a training set is required which contains vector pairs describing the desired mapping. Such a training set could be provided off line (but may be continued on line), as follows. The controller and plant model are set up in closed loop and an initial set of gains chosen. A step in demand is made and the resulting labels recorded as L_0. The gains are perturbed and labels generated for the resulting step response. The label error L_e is given by $L_0 - L$. L_e and the negated gain perturbation, $-\Delta K$, provide one training pair. However problems arise in ensuring that an adequate range of label errors is generated in the training set. It is the label error vectors that are mapped to gain changes in the tuning module but unfortunately these cannot be directly manipulated for training set generation. One solution to this problem is to increase the ΔK space sufficiently to ensure that all label error vectors likely to arise in practice are represented; this is a common problem in the realm of neural network control [11]. A potentially more serious problem is that a certain class of label error vectors may never arise in training and thus their

Fig. 7.3 Simulation results.

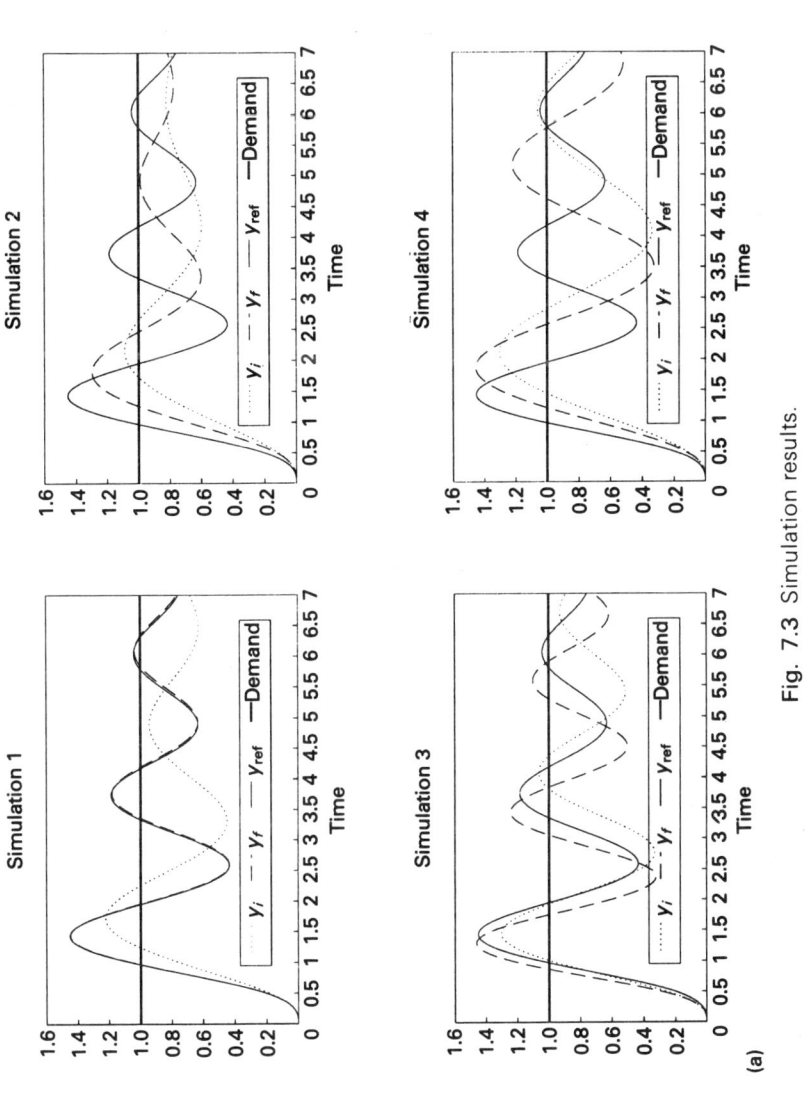

Fig. 7.3 – contd.

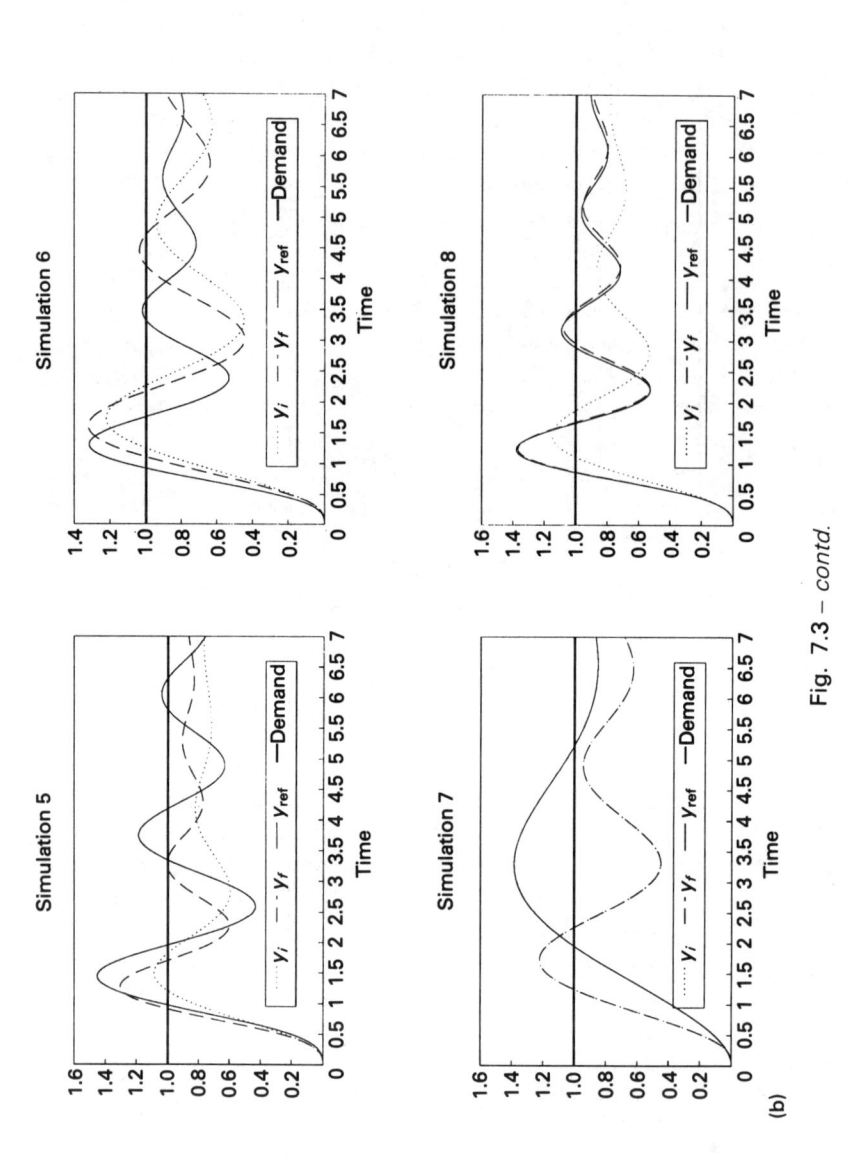

occurrence in operation will stimulate a zero output from the net. This is demonstrated in simulation 7, see Fig. 7.3.

It is noted that the efficiency of the resulting CMAC tuning module is dependent on a realistic choice of reference model, i.e. one that is achievable (or nearly so) with the given controller and gain ranges. For example we cannot expect zero overshoot and zero SSE for a second order system under proportional control. Proportional control can only provide a curve in the label space parametrized by P, the proportional gain. This may be represented as $L(P)$. Thus only a small set of the possible label error combinations are encountered during training. Indeed, after investigation, it was found that only 25% of the memory locations had been initialized during training for the simulations discussed in section 7.5.

It is reasonable to make the comment that this is a brute force, rather than an intelligent, approach. However, training may be viewed as a method of providing the net with experience; it learns to associate gain changes with response changes, in perhaps the same way an instrument engineer might. It is expected that the iterative tuning provided by the trained net will provide a robust search, may be updated on line and, furthermore, will provide a fast input–output response. Early results, although somewhat contrived, are encouraging. These are described in the results section.

7.5 RESULTS

Simulations were carried out using a digitized linear model and control system. The plant's running speed was chosen as 500 Hz and the control system's was 50 Hz. In this way the simulations approximated a digital control system operating on an analogue plant. The plant chosen was a second-order linear system with arbitrarily chosen parameters ($\omega_n = 1.0$, $\zeta = 0.3$). The reference model was chosen as the plant in closed loop proportional control with a gain of 6.5. Seventeen training pairs were required to demonstrate the action of the tuning module. A gain of 5 was chosen as the bench mark gain, P_0, and was varied in the range [1.0, 9.0] in steps of 0.5. The learning rate decay parameter was chosen to be 50.0 and ten presentations of the training set were made. Eight simulations were carried out as in Table 7.1 where P_i is the initial controller gain. The label error vector was chosen as (OS%, T_r, SSE%) and is defined in (7.8) where FV is the final value estimate and the demand is a unit step.

$$\text{OS\%} = 100(1 - \text{peak value}), \qquad T_r = t_2 - t_1$$

where

$$y(t_1) = 0.1\text{FV}, y(t_2) = 0.9\text{FV}, \text{SSE\%} = 100(1 - \text{FV}). \tag{7.8}$$

The results are illustrated graphically in Fig. 7.3. The gain values at each iteration are recorded in Table 7.2 and the initial and final label errors in

98 A label-driven CMAC intelligent control strategy

Table 7.1 Simulation details

Number	P_i	ω_n^p, ζ^p	P_{ref}	$\omega_n^{ref}, \zeta^{ref}$
1	3.0	1.0, 0.3	6.5	1.0, 0.3
2	3.0	0.8, 0.5	6.5	1.0, 0.3
3	3.0	1.2, 0.2	6.5	1.0, 0.3
4	3.0	0.8, 0.2	6.5	1.0, 0.3
5	3.0	1.2, 0.5	6.5	1.0, 0.3
6	3.0	1.0, 0.3	5.0	1.2, 0.4
7	3.0	1.0, 0.3	OL	1.0, 0.3
8	3.0	1.2, 0.4	6.5	1.2, 0.4

Table 7.2 Progression of K_p during tuning for eight simulations

Iteration	1	2	3	4	5	6	7	8
0	3.00	3.00	3.00	3.00	3.00	3.00	3.00	3.00
1	5.30	5.59	4.56	5.57	4.64	5.30	–	5.45
2	6.58	5.62	5.10	5.21	5.87	5.13	–	6.19
3	6.44	–	4.99	5.04	5.98	4.97	–	–
4	6.40	–	–	4.90	–	4.81	–	–
5	6.38	–	–	4.77	–	4.65	–	–
6	–	–	–	4.68	–	4.49	–	–
7	–	–	–	–	–	4.33	–	–
8	–	–	–	–	–	4.18	–	–
9	–	–	–	–	–	4.03	–	–
10	–	–	–	–	–	3.98	–	–

Table 7.3. Convergence was deemed to have occurred when the gain change, ΔP, fell below a predetermined threshold. This was set at 0.02 throughout the simulations. For the first simulation the plant and reference models have the same parameters, hence zero label errors are obtainable. The tuning process is completed in five iterations and 95% of the absolute change takes place in the first two iterations. Label errors are greatly reduced for instances in which zero errors are unobtainable, i.e. when plant and reference models differ due to small changes in the plant parameters. This is

Table 7.3 Initial and final label errors for eight simulations

Simulation	Initial L_e	Final L_e
1	23.09, −0.10, −8.90	0.34, 0.00, −0.42
2	36.50, −0.34, −12.61	15.41, −0.22, 0.85
3	14.88, 0.08, −11.66	−1.03, 0.10, −0.85
4	14.91, −0.24, −13.85	−0.30, −0.18, −2.50
5	36.46, 0.00, −10.40	14.96, 0.08, 0.70
6	9.64, −0.18, −7.24	−0.45, −0.14, −2.28
7	16.26, 0.88, −25.74	16.26, 0.88, −25.74
8	22.95, −0.08, −12.15	1.02, −0.02, −0.33

demonstrated in simulations 2–5. For simulation 6 the reference model labels were generated using different parameters to those of the training data generation plant model. This caused slower convergence although 65% of the absolute change took place in the first iteration. For simulation 7 unrealistic labels were generated from the given model's open loop response, notably zero SSE and a long rise time. The net has no experience of the label errors generated and consequently produces a zero output. The final simulation demonstrates the importance of the mapping's character, or its topology, rather than exact numerical values. Zero label errors are achievable and are approximately achieved in two iterations.

By examination of the training data it is possible to discern the character of the desired mapping, i.e. the octants through which $L_e(\Delta P)$ passes. This parametrized curve must always pass through the origin. Its character is represented thus;

L_e		ΔP
+, −, −	→	+
0, 0, 0	→	0
−, +, +	→	−

Slices through the net's output for three values of rise time error are given in Fig. 7.4. The uninitialized quadrants are clearly evident and correspond to overshoot-SSE error combinations of +, + and −, −.

In all simulations, other than the first and seventh, label error combinations other than the above are eventually encountered stimulating a small output from the net (≤ 0.02). Simulations 2 and 5 are of particular interest with regard to label error combinations. The initial combination for simulation 2 is of the form (+, −, −). The tuning module increases the gain until a combination of the form (+, −, +) is encountered. The net produces a small or zero output and tuning stops. It is evident that under proportional control this methodology cannot reduce the label error vector as a whole, i.e. it cannot minimize $\|L_e\|$ for some norm $\|\cdot\|$. In the first simulation the above error combinations are maintained as zero label errors are achievable and the plant parameters are as for training. In the seventh simulation a rise time error of 0.9 is encountered. However during training it does not exceed 0.02 due to actuator rate saturation. Hence a small or zero output is produced, see Fig. 7.4(c).

7.6 CONCLUSIONS AND FURTHER WORK

The tuning of a proportional controller via a CMAC tuning module has been investigated through simulation. The simulations carried out were simple and artificial but demonstrate that this methodology requires further research. The module achieved close matching between the reference model

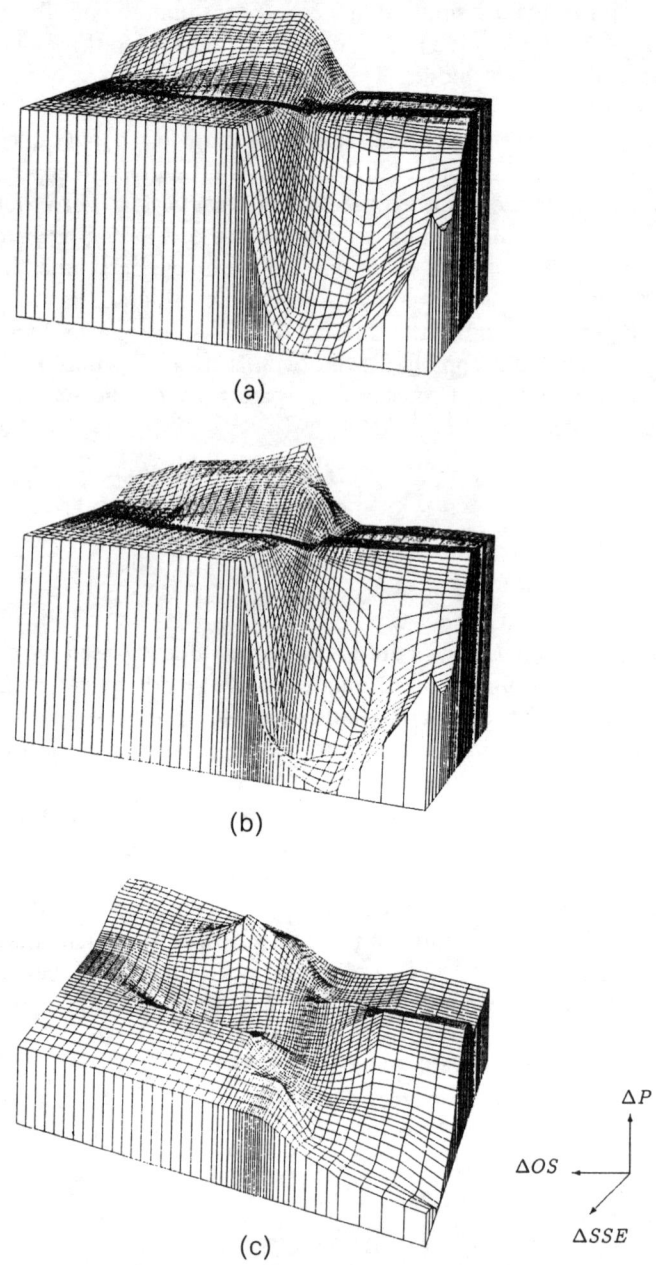

100 A label-driven CMAC intelligent control strategy

Fig. 7.4 (a) rise time error = −0.1; (b) rise time error = 0.02; (c) rise time error = 0.9. In (a) and (b) the output, ΔP, is in the range (−7.0, 3.0). In (c) the rise time error of 0.9 is outside the training range and thus the output is small or zero. This is to be expected as CMAC's strengths are local generalization and interpolation.

Conclusions and further work

Table 7.4 Net description summary

Network dimension is 3
Generalization parameter is 7
Output dimension is 1
Knot matrix, a ragged matrix where $k(i, j) = i$th knot on the jth axis.

$$\begin{pmatrix} -40.00 & -1.00 & -50.00 \\ -20.00 & -0.30 & -40.00 \\ -10.00 & -0.20 & -30.00 \\ -5.00 & -0.10 & -20.00 \\ -2.00 & 0.00 & -15.00 \\ 0.00 & 0.10 & -10.00 \\ 2.00 & 0.20 & -7.00 \\ 5.00 & 0.30 & -5.00 \\ 10.00 & 1.00 & -3.00 \\ 20.00 & & -2.00 \\ 30.00 & & -1.00 \\ 40.00 & & 0.00 \\ 50.00 & & 1.00 \\ 60.00 & & 2.00 \\ 80.00 & & 3.00 \\ & & 5.00 \\ & & 5.00 \\ & & 7.00 \\ & & 10.00 \\ & & 15.00 \end{pmatrix}$$

Offset matrix, $d(i, j) =$ the offset of the ith subset on the jth axis

$$\begin{pmatrix} 1 & 2 & 3 \\ 2 & 4 & 6 \\ 3 & 6 & 2 \\ 4 & 1 & 5 \\ 5 & 3 & 1 \\ 6 & 5 & 4 \\ 7 & 7 & 7 \end{pmatrix}$$

Combining rule is: product
Univariate basis function shape is: hat
Update rule is: NLMS
Learning rate mode is: variable
Variable learning rate parameter = 50.00
Total memory locations = 138

and plant step responses in the sense that initial label errors were greatly reduced. The fact that these simulation results were achieved with so few training vector pairs indicates that there exists the potential for more sophisticated tuning tasks.

Work is now being undertaken to extend this approach to PID control under the effects of sensor noise. Coping with sensor noise will entail filtering the response signal (derivative control requires this anyway) and the writing of robust label computation routines. Under PID control the set of possible label combinations will be much richer, parametrically a function of the three PID gains, $L(K)$. It is expected that this will allow a larger set of reference model labels to be used for tuning. However, this may be offset to some extent as a larger label set will be required in order to describe the more varied response shapes. This will be especially evident for control problems involving high-order dynamics which are not dominant second order. The effects of K_0's choice must be investigated. It seems likely that it must be in the right 'ball park', so to speak.

There is scope for extending the approach to cope with a non-linear plant control problem. The obvious solution is to design tuning modules for an array of operating points. These may then be assigned to operating regions or perhaps splined together. The simulations have demonstrated that the tuning module can cope with non-linearities that may be modelled by a linear structured plant subject to parametric variation. Hence each tuning module can be expected to perform well in its specified operating region.

It is important to note that in the simulations described the tuning module does not have explicit knowledge of the plant. The second-order properties of the plant manifest themselves in the label and reference model choices. The labels are chosen in order to best describe the salient features of the response and are in no way restricted to processes that are dominant second order. As already mentioned, a realistic choice of reference model must be made, which is in agreement with common sense. In the light of these observations it is recognized that conditions on these parameters must be determined which ensure a successful control system.

REFERENCES

1. Albus, J.S. (1975) A new approach to manipulator control: the Cerebellar Model Articulation Controller (CMAC). *Trans. ASME J. Dyn. Sys., Mea. & Control*, **97**, 220–33.
2. Albus, J.S. (1979) A model of the brain for robot control. Part 2. *Byte*, **July**, 54–95.
3. Albus, J.S. (1979) Mechanisms of planning and problem solving in the brain. *Mathematical Biosciences*, **45**, 247–93.
4. Miller, W.T., Glanz, F.H. and Kraft, L.G. (1990) CMAC: an associative neural network alternative to backpropagation. *Proc of the IEEE*, **78**(10), 1561–7.
5. Kraft, L.G. and Campagna, D.P. (1990) A comparison between CMAC neural network control and two traditional adaptive control systems. *IEEE Control Sytems Mag.*, **10**(3), 36–43.
6. Harris, C.J. (1992) *Intelligent control: Aspects of fuzzy logic and neural nets.* World Scientific Press, Singapore.
7. Parks, P.C. and Militzer, J. (1991) Improved allocation of weights for associative memory storage in learning control systems. *Proc. 1st IFAC Symposium on Design methods of control systems*, Zurich, **2**, 777–82.
8. Lawrence, A.J. and Harris, C.J. (1992) CMAC and its modifications. Internal report, Aero and Astro Dept., Southampton University, UK.

9. Fraser, R. (1989) Application of CMAC algorithm to control of dynamical systems. Project No. 000927, Aero and Astro Dept., Southampton University, UK.
10. Kumar, S.S. and Guez, A. (1991) ART based adaptive pole placement for neurocontrollers. *Neural Networks*, **4**, 319–35.
11. Psaltis, D. *et al.* (1988) A multilayered neural network controller. *IEEE Control Systems Mag.*, **8**(2), 17–21.

Neural network controller for depth of anaesthesia

8

D.A. Linkens and H.U. Rehman

8.1 INTRODUCTION

Sleep and anaesthesia are both states of unresponsiveness which vary in depth. Sleep is natural, healthy and has a circadian rhythm (a rhythm of an approximately twenty-four-hour period) [1]. Anaesthesia is an artificial state maintained by the continuing presence of chemical (anaesthetic) agents in the brain. Using anaesthetic agents there are several complications and side effects, which may cause death under extreme conditions. The first death from the use of chloroform was recorded in 1848 [2].

Anaesthetic agents affect the respiratory system, cardiovascular system, central nervous system and muscles [3]. The effect on all these systems can provide a monitoring system for the safe use of the anaesthetic agents. Since there is no direct method of measuring the depth of anaesthesia, various techniques are used to suggest the safe amount of anaesthetic drug, considering the condition of an individual patient. Monitoring of electroencephalogram (EEG) signals, colour of skin, pupil size, patient movement and other clinical signs such as heart rate (HR), respiration rate (RR), systolic arterial pressure (SAP) and mean arterial pressure are used by anaesthetists to determine the depth of anaesthesia [4]. Usually, anaesthetists decide the correct dose of anaesthetic agent on the basis of their own experience and skill. However, the clinical signs may be misleading for a variety of reasons, and this may cause an incorrect dosage to be supplied. Further, there is a considerable variation in the importance and interpretation applied to each clinical sign.

8.2 BACKGROUND AND HISTORY

Studies have been carried out by various workers [5, 6, 7] to control the depth of anaesthesia using both open-loop and closed-loop techniques [8]. A recent development is a computer-based on-line expert system called RESAC (realtime expert system for advice and control) [9]. It comprises a rule-based backward-chaining inference engine with about 400 rules, and makes use of fuzzy-logic and Bayesian reasoning. This size of rule-base appears to be at the threshold of realistic rule acquisition, editing, debugging and verification of the knowledge base. Neither rule-based nor frame-based artificial intelligence paradigms are easy to manage with this size of knowledge domain.

In contrast, neural networks (NN) offer a better possibility of rapid knowledge acquisition via their self-organizing learning properties. They have the ability to learn in those cases where it is possible to specify the inputs and outputs but difficult to define the relationship between them. They are also tolerant to noise in the input data. These attributes of NN are suitable for the case of anaesthesia because the relationships between clinical signs are not clear. Using this technique there is no problem associated with the number of rules as the relationships are developed by NN itself. No other work is known to have been reported on the application of NN to the determination of the depth of anaesthesia.

8.3 PRESENT WORK AND DISCUSSION

During several surgical operations, patients' data such as HR, SAP, RR, percentage of anaesthetic agent and temperature were logged into a computer. These data were input to RESAC which gave certainty values from -500 to 500 for three states of anaesthesia and a suggested dosage ranging from 0% to 5%. The intention was to get smooth data, and to compare the NN results with those of RESAC. It was observed in the initial stages of the study that NN took less time and gave better results if data were smooth rather than having big transitions. The certainty values of the three states of anaesthesia and the percentage dosage have been used as target outputs in this study.

The NN back-propagation learning paradigm [10] has been used to train the program with inputs of age, weight, gender, HR, SAP and RR. After a series of training and testing sessions, results were compared with those of RESAC and found to be very promising. In subsequent trainings, time delays in the anaesthetic dose were introduced, as the effect of a previous dose remains for some time. Comparison of results, with one time delay (Fig. 8.1(a) to (d)) and two time delays (Fig. 8.2(a) to (d)) shows that the former gave a better result than the latter. This is also in accordance with the pharmacology of the anaesthetic agent (isoflurane was used in all

106 Controller for depth of anaesthesia

(a)

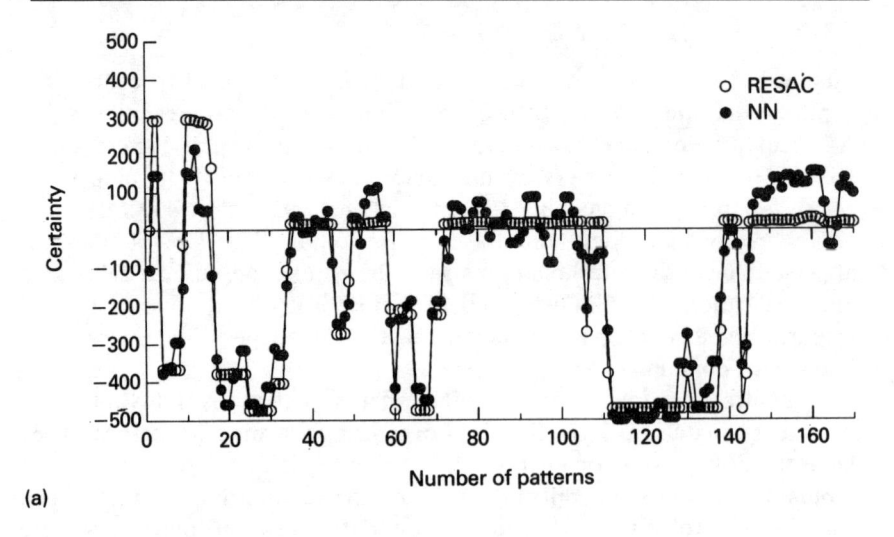

(b)

Fig. 8.1 The graphs between outputs of RESAC and NN with one time delay in dosage: (a) anaesthetic OK; (b) anaesthetic light; (c) anaesthetic deep; and (d) dosage.

operations) which suggests that the uptake and the elimination of isoflurane is quicker than the other anaesthetic agents like enflurane [11]. Also delta target values (delta target = target − original) for HR, SAP and RR, desired by anaesthetists for individual patients undergoing surgery, were included as inputs for the training of the program.

Successive trainings were carried out with ten inputs and four outputs, selecting different numbers of hidden units in the network. The minimum

(c)

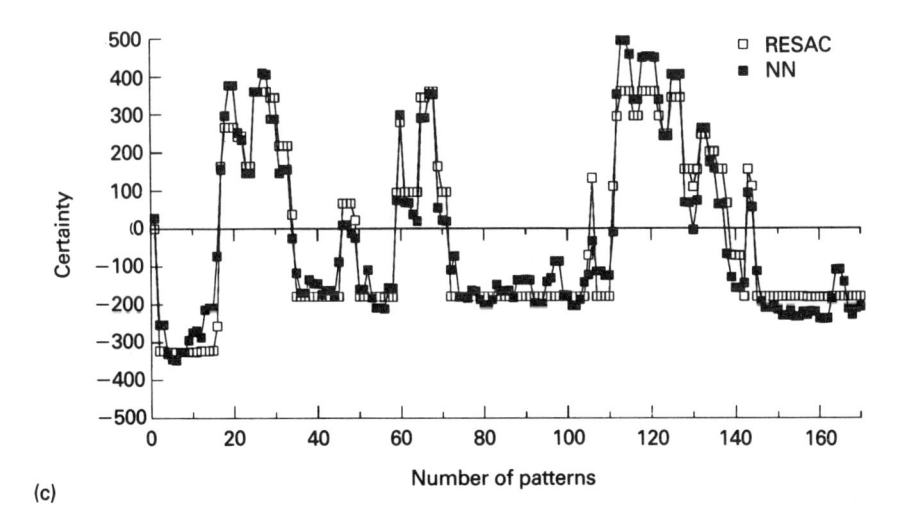

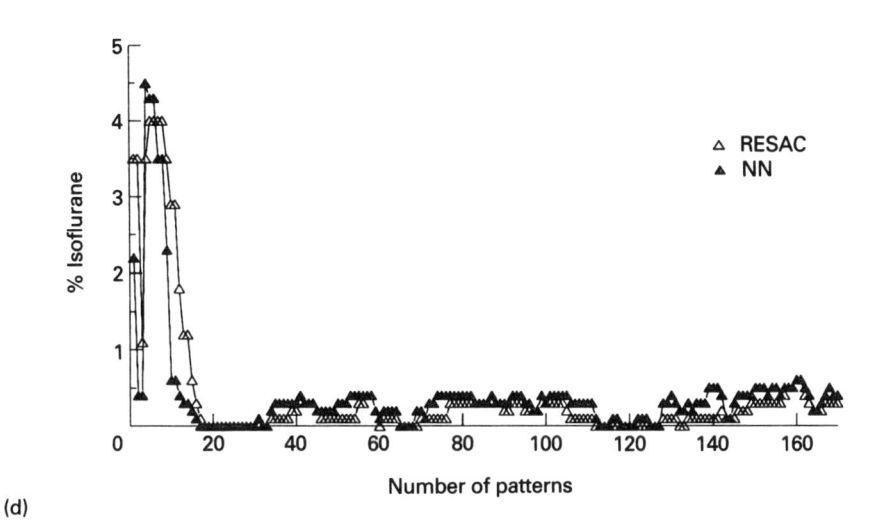

(d)

Fig. 8.1 – *contd.*

value of the total sum of squares [10] was obtained with three hidden units. For initializing the training network, random biases and weights were selected. Also the learning rate and momentum term for upgrading the weights were gradually increased as the training proceeded. This gradual increase in learning rate and momentum term helps to converge the system more rapidly. After the successful training and testing, a computer program called ANNAD (artificial neural network for anaesthetic dose) was developed, using the weights from the above training runs. ANNAD suggests the

108 Controller for depth of anaesthesia

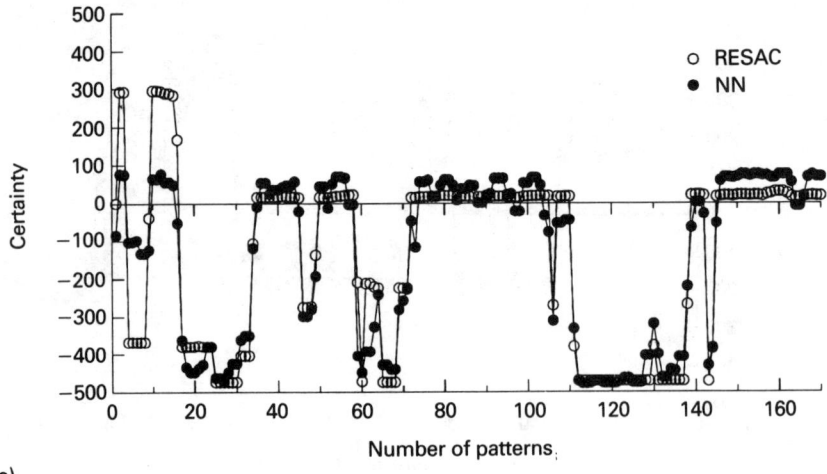

(a)

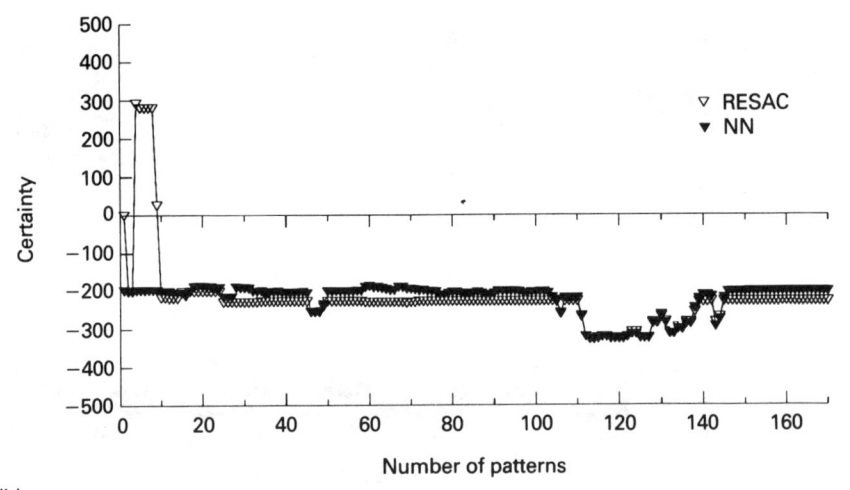

(b)

Fig. 8.2 The graphs between outputs of RESAC and NN with two time delays in dosage: (a) anaesthetic OK; (b) anaesthetic light; (c) anaesthetic deep; and (d) dosage.

dosage of anaesthetic agent and three states of anaesthesia, after receiving time-related inputs. In experiments conducted so far the three states of anaesthesia and suggested dosage by ANNAD have been almost the same as that suggested by RESAC. Figure 8.3 shows the states of anaesthesia and suggested dose by RESAC and ANNAD for a particular patient.

An interesting feature of ANNAD is demonstrated in Fig. 8.4(a) to (d), where two different sets of target values of HR, SAP and RR for another

(c)

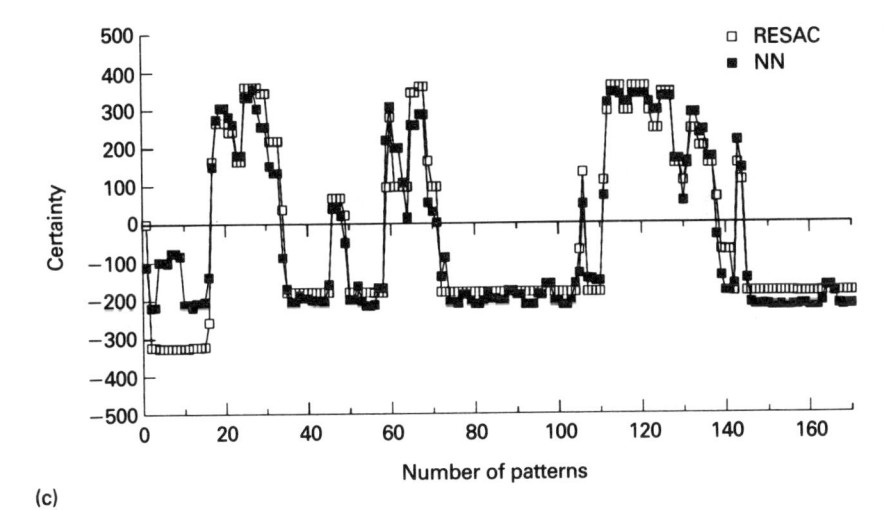

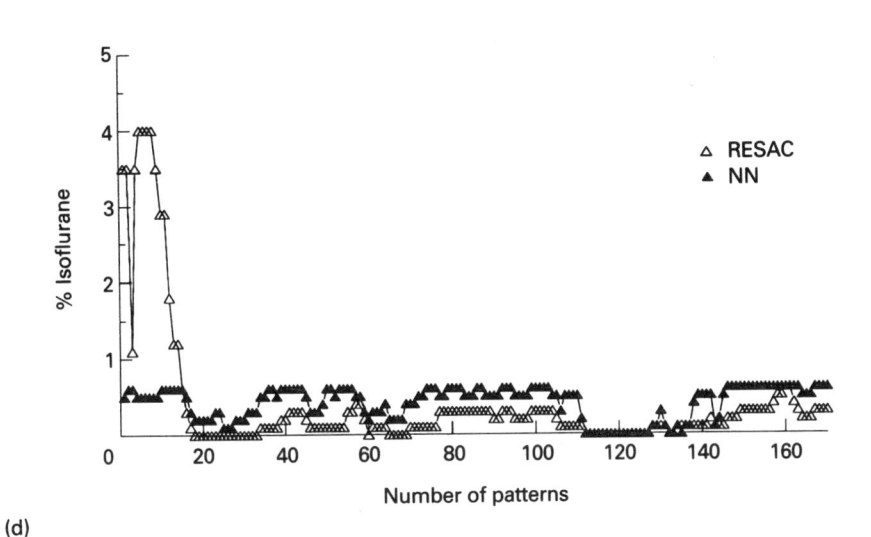

(d)

Fig. 8.2 – contd.

patient are presented to ANNAD, and graphs are drawn against the actual outputs by RESAC. It can be seen that the outputs by ANNAD for each set of targets are different, which ultimately suggests that ANNAD does take account of target values.

Due to the clinical limitations the above program cannot be tested during surgical operations at this stage. Thus, ANNAD has been connected to another program that models the patient, obtained via regression analysis on actual clinical data. Closed-loop studies can be performed with these two

110 Controller for depth of anaesthesia

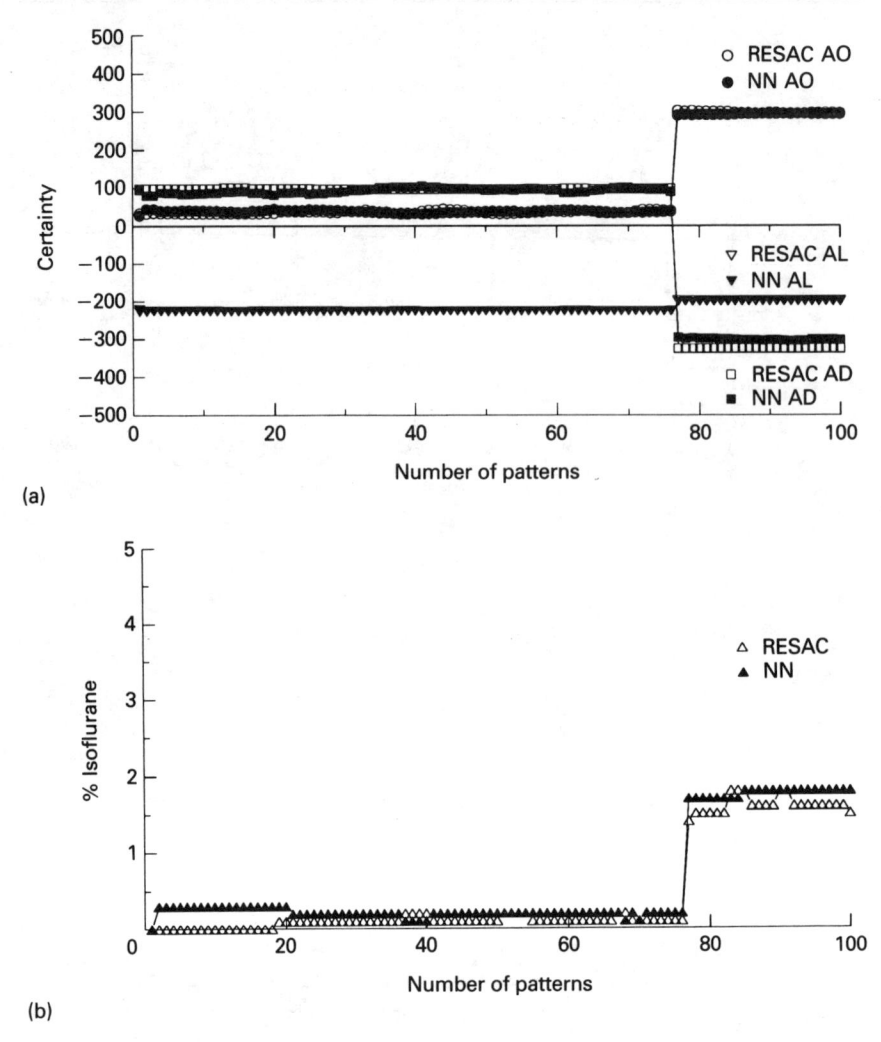

Fig. 8.3 Outputs of RESAC and ANNAD; (a) three states of anaesthesia, i.e. anaesthetic OK (AO), anaesthetic light (AL) and anaesthetic deep (AD) and (b) percentage dosage.

programs since the patient simulator is able to supply clinical signs to ANNAD which in turn supplies the suggested doses influencing the patient model. Figure 8.5 represents the structure of closed-loop control of ANNAD and the patient model. ANNAD has been tested with various patient models to check its compatibility. Figure 8.6(a) and (b) show the percentage of suggested dose together with the clinical signs supplied by the patient model. The system tends to an equilibrium which depends upon target values and initial data of the patient.

(a)

(b)

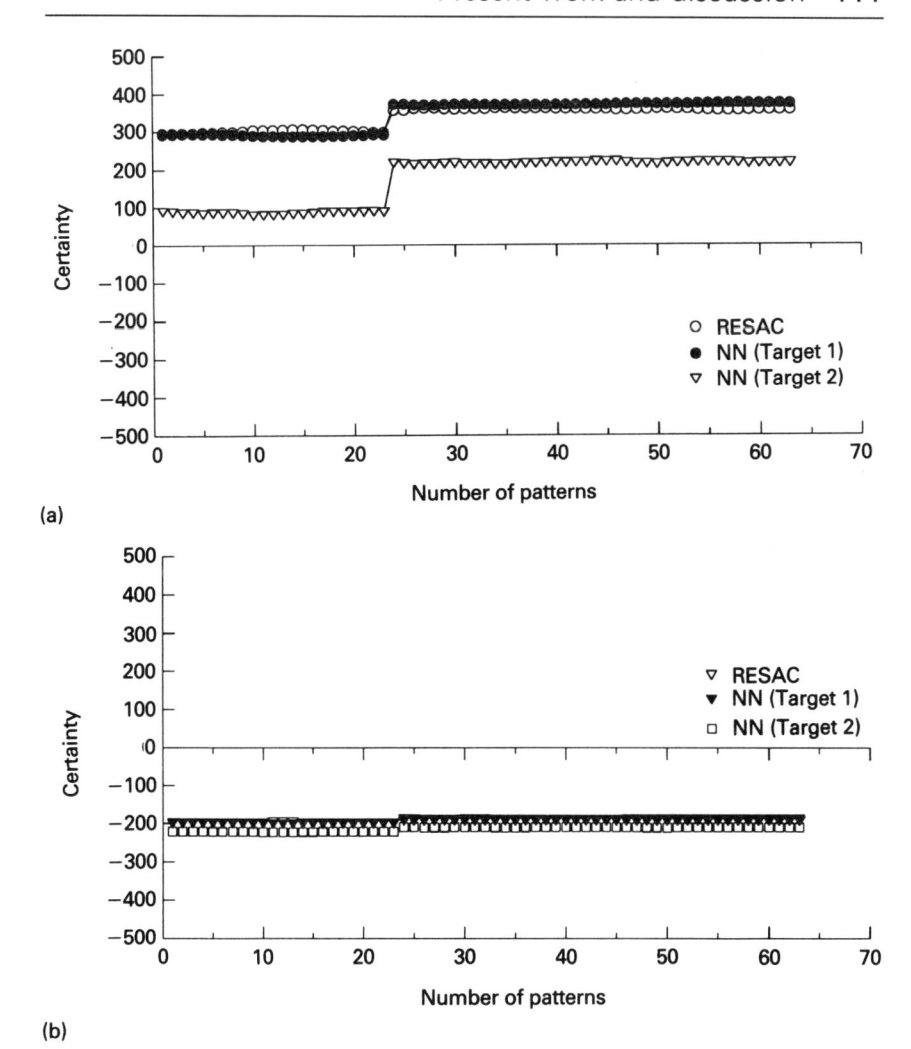

Fig. 8.4 Comparison of outputs from RESAC and ANNAD with two sets of target values. For the first set, target HR=80, target SAP=110 and target RR=18, while for the second, target HR=85, target, SAP=120 and target RR=20: (a) anaesthetic OK; (b) anaesthetic light; (c) anaesthetic deep; and (d) dosage.

112 Controller for depth of anaesthesia

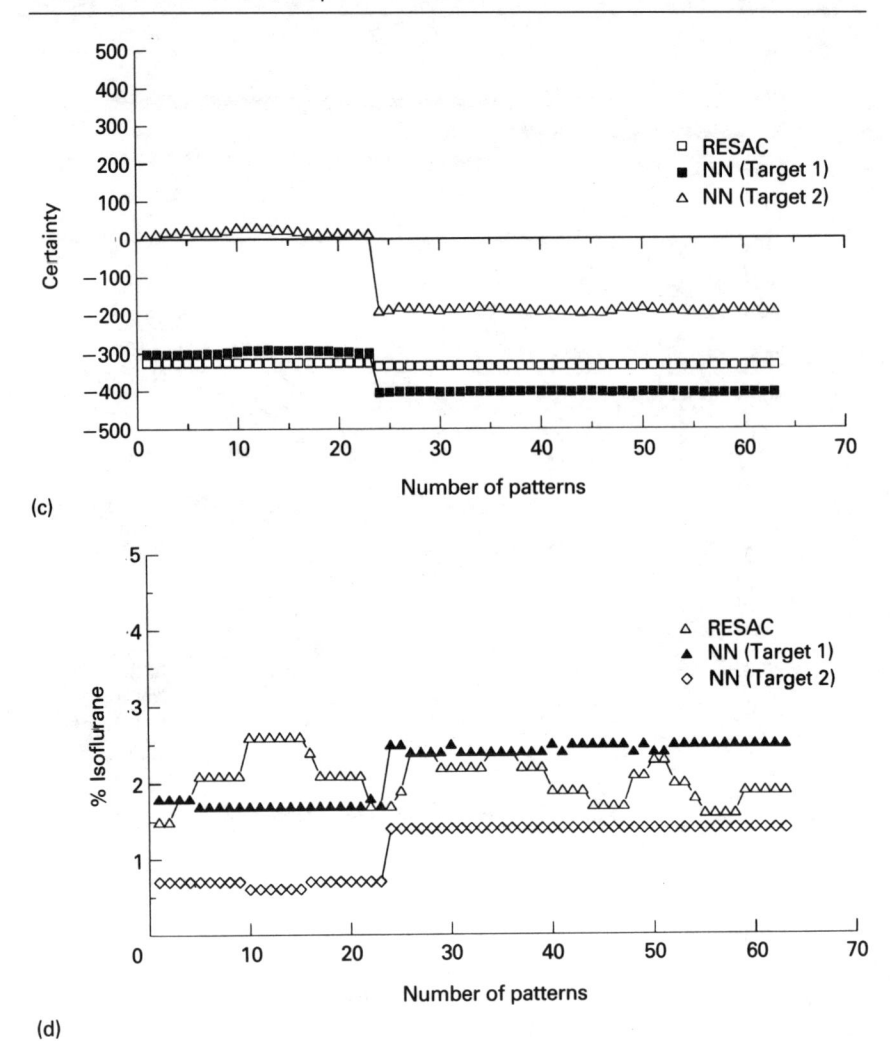

(c)

(d)

Fig. 8.4 – *contd.*

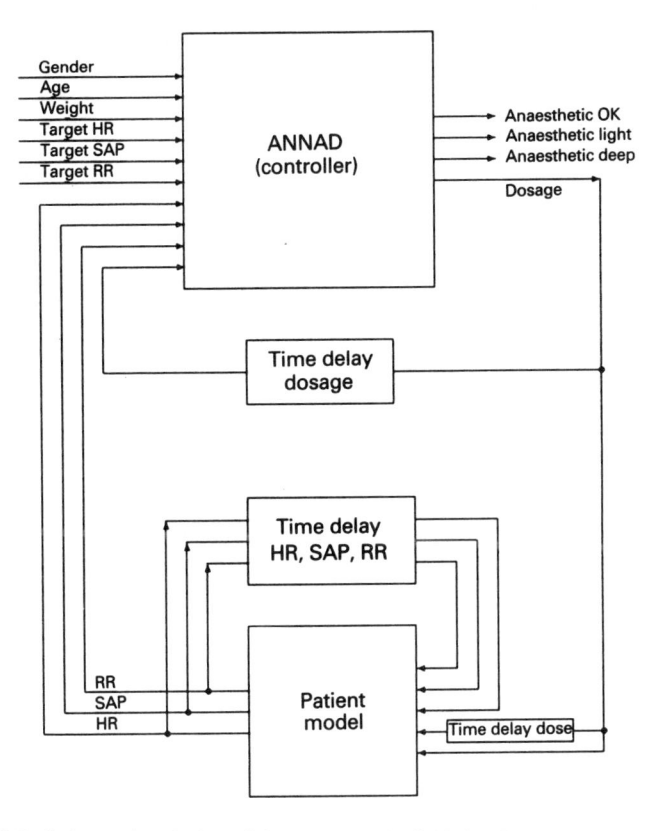

Fig. 8.5 Schematic of closed-loop control of ANNAD and patient model.

114 Controller for depth of anaesthesia

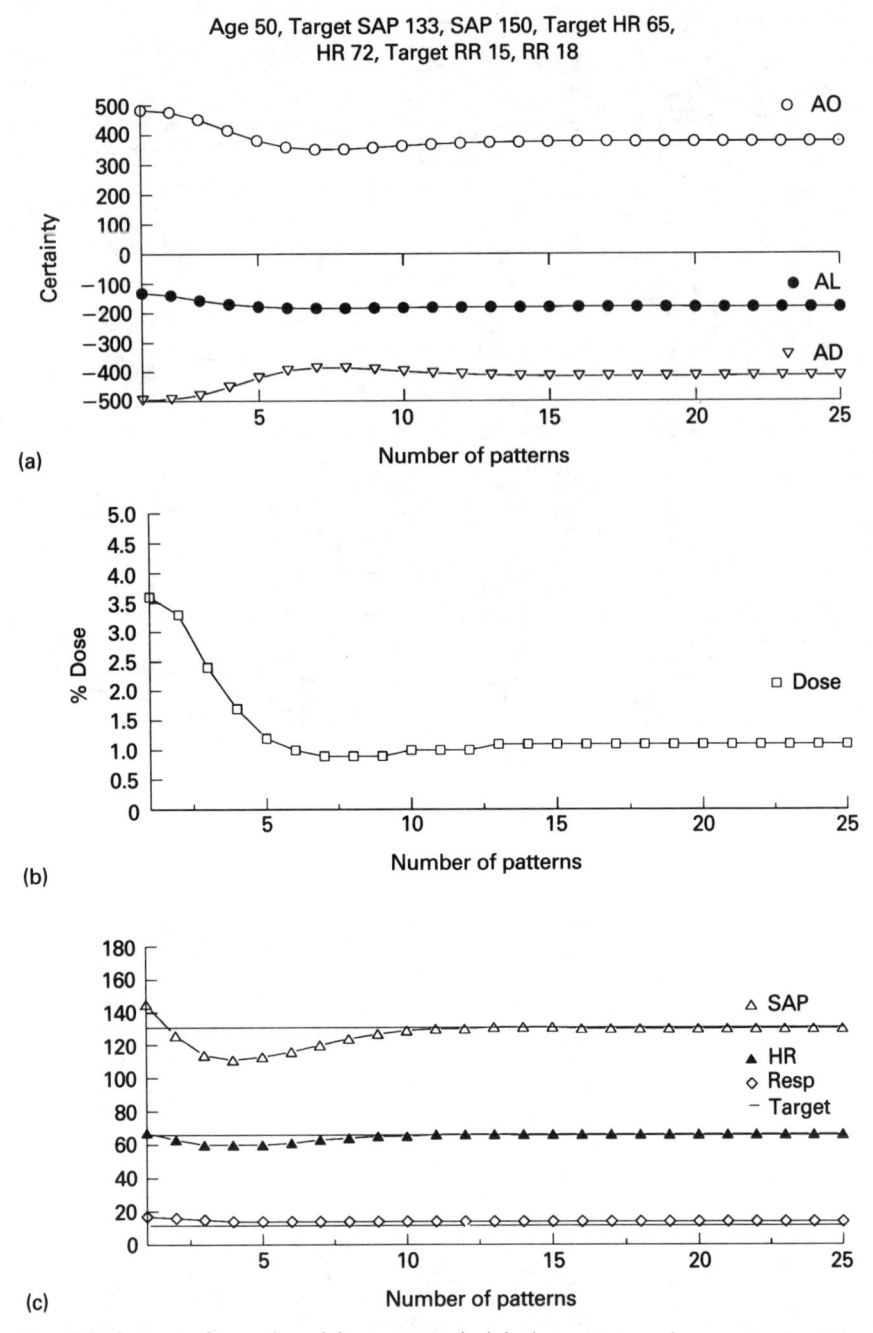

Fig. 8.6 Outputs from closed-loop control: (a) three states of anaesthesia; (b) drug dose from ANNAD; and (c) HR, SAP and RR from patient model.

8.4 CONCLUSIONS

The results demonstrate the ability of an NN to replicate advice from RESAC, and also the control performance obtainable when this is coupled to a patient simulator. The patient model used in this study is obtained via linear regression. However, NN patient models are also under development. To compare the performance of ANNAD, a multivariable non-linear regression controller is also under consideration, which will also be connected to patient models in a closed loop.

In actual clinical trials ANNAD may be interfaced to work with a Dinamap blood-pressure monitor and other instrumentation to supply necessary on-line information directly.

REFERENCES

1. Oswald, I. The physiology and pharmacology of sleep. In *Anaesthesia*, Vol. 1, (ed. W.S. Nimmo and G. Smith). Blackwell Scientific Publications, Oxford, pp. 3–9.
2. Atkinson, R.S., Rushman, G.B. and Lee, A. (1987) *A Synopsis of Anaesthesia*, 10th edn. Wright, Bristol.
3. Vickers, M.D. and Schnieden, H. (1984) *Drugs in Anaesthesia Practice*. Butterworth, Sevenoaks.
4. Linkens, D.A., Greenhow, S.G. and Asbury, A.J. (1986) An expert system for the control of depth of anaesthesia. *Biomed. Meas. Infor. Contr.*, 1(1), 223–8.
5. Chilcoat, R.T., Lunn, J.N. and Mapleson, W.W. (1984) Computer assistance in the control of depth of anaesthesia. *British Journal of Anaesthesia*, 56, 1417–31.
6. Robb, H.M., Asbury, A.J., Gray, W.M. and Linkens, D.A. (1991) Towards a standardized anaesthetic state using enflurane and morphine. *British Journal of Anaesthesia*, 66, 358–64.
7. Greenhow, S.G. (1990) A knowledge based system for the control of depth of anaesthesia. Ph.D. Thesis, University of Sheffield, UK.
8. Linkens, D.A. and Hacisalihzade, S.S. (1990) Computer control system and pharmacological drug administration: a survey. *Journal of Medical Engineering and Technology*, 14(2), 41–54.
9. Linkens, D.A., Greenhow, S.G. and Asbury, A.J. (1990) Clinical trials with the anaesthetic expert advisor RESAC. *Expert Systems in Medicine*, 6th Annual Meeting, London, pp. 11–18.
10. Rumelhart, D.E., Hinton, G.E., and Williams, R.J. (1986) Learning internal representations by error propagation. In *Parallel Distributed Processing: Explorations in the Microstructure of Cognition*, Vol. 1 (ed. D.E. Rumelhart, J.L. McClelland and the PDP Research Group). MIT Press, Cambridge, Mass.
11. Stoelting, R.K. and Miller, R.D. (1989) *Basics of Anesthesia*, Churchill Livingstone, New York.

Index

Entries in **bold** refer to figures

Activation functions
  hard limiter 3–4, **5**
  sigmoid 3–4, **5**, 37
    polynomial approximation 35, 41–2
  tanh 3–4
  threshold 3–4
Adaptive critic, *see* Control
Adaptive networks 44
ANNAD, *see* Artificial neural network for anaesthetic dose
AR, *see* Auto-regressive models
Artificial neural network for anaesthetic dose 107–10
Artificial neuron model 3
  see also McCulloch and Pitts neuron model
Associative layer 11
  see also Hidden layers
Associative memory networks 90
Auto-regressive models
  linear systems 55–6
    results 56–9
  non-linear systems 59–60
    results 60–1
Axon 1

Biological neuron **2**
Back-error propagation 13
  see also Back-propagation algorithm
Back-propagation algorithm 25, 27, 38, 79–80
  see also Training algorithms
Basis function 90–1
BEP, *see* Back-propagation algorithm
BOXES method 13
B-splines 90

Cerebellar model articulation controller
  algorithm 90–2
  control scheme 13–14, 92–4
  possible architecture **93**
  tuning module 94–7
  training 92
  results 97–9
CMAC, *see* cerebellar model articulation controller
Conjugate gradient method 38
Context units 26
Control
  adaptive 89–90
  adaptive critic 19
  of anaesthesia 104–5
  copy control 19
  differentiating the model 21
  direct inverse control 75
  internal model control 75
  model reference control 21
  predictive control 15, 20
    strategy 84–5
    results 85–7
Copy control, *see* Control
Correlation analysis to reduce network

118 Index

topology 63–5
Correlation training 12
Cost function 5
Credit assignment problem 12–13

Data conditioning
  non-linear discriminant surfaces 16
  scaling 62
  single-node data mapping 76–7
  spread encoding 16, 76–7
Decorrelation techniques 18
Dendrites 1
Differentiating the model, *see* Control

Elman nets
  basic network 26–7
  modified network 27–9
Estimation 13–18

FANN, *see* Feedforward neural networks
Feedforward neural networks 25, 37–9, 61–2
  see also Neural network
Fermentation processes 53–5
Fuzzy logic 90

General-purpose algorithm 14
Gradient descent 12, 13
  see also training algorithms

Hashing methods 13, 17
Hidden layers 4
Hopfield network
  structure 1, 3, 5
  training 5

Image processing 6, 7
Input units 4, 26

Kalman filter estimator 56
Kohonen network
  adaptation zone 6
  structure 3, 6–7
  training 6

Linear discriminants 11

Matrix associative memories 12
Matrix multiplier 11
McCulloch and Pitts neuron model **3**

Melt flow index estimation 48–50
Michie and Chambers 13
MLP, *see* Multi-layer perceptron
Model reference control, *see* Control
Multi-layer perceptron
  back-propagation algorithm 4
  cost function 5
  feedforward network 4
  network training 4
  structure 3–4, 15

NARX model, *see* Non-linear, autoregressive exogenous model
Neural network
  as a non-linear adaptive filter 10
  attributes 2
  biological aspects 1
  feedforward 25, 37–39, 61–2
  geneology **14**
  model structure 80, 81
  predictor structure 80
Neural predictive control
  results 85–7
  strategy 84–5
  structure **80**
Nodes
  context 26
  hidden 27
  input 26
  output 26
Non-linear adaptive filters 10
Non-linear, auto-regressive exogenous model 79
Non-linear discriminant surface 16
  see also Data conditioning

Optimum topology selection 42–3
Output units 4, 26

Pattern recognition 6, 7
PCA, *see* Principal component analysis
Perceptron 10
  see also Multi-layer perceptron
Plant estimators 15
Plant inverse 22
Predictive control, *see* Control
Polynomial-based non-linear processing functions 35
Post-pruning processes 64–6
Principal component analysis
  application 40–2

Index

theory 39–40
Product conversion estimation 45–8

Radial basis functions 15, 90
Recursive networks 16
Recurrent networks 25
Reinforcement learning 12
Realtime expert system for advice and control 105, 108–9
RESAC, *see* Realtime expert system for advice and control

Self-tuning networks 44
Sensory layer 10–11
Single-node data mapping 76–7
  see also Data conditioning
SNDM, *see* Single-node data mapping
Spread encoding
  sliding Gaussian distribution **76**
  technique 76–7

see also Data conditioning
State estimation 61–8
Steepest descent method 38
  see also Back-propagation algorithm
Stop-learning criterion 63

Training algorithms
  back-propagation 38
  gradient descent 12, 13
  least mean square 92
  speed 36
  steepest descent 38
  supervised 4–5

Units, *see* Nodes

Widrow and Hoff 11

XOR problem 11